Developmental Arithmetic — A Computational Review

Sixth Edition

James C. Curl

**Modesto Junior College
Modesto, California**

PEARSON
Custom Publishing

Copyright © 2007, 2003, 1992, 1985, 1980 by James C. Curl

All rights reserved.

Copy Editor: Heather Curl

Illustrator and Typographer: Joshua Delahunty

Permission in writing must be obtained from the publisher before any part of this work may be reproduced or transmitted in any form or by any means, electronic or mechanical, including photocopying and recording, or by any information storage or retrieval system.

All trademarks, service marks, registered trademarks, and registered service marks are the property of their respective owners and are used herein for identification purposes only.

Printed in the United States of America

18 17 16 15

ISBN 0-536-45156-7

2007360080

JK

Please visit our web site at *www.pearsoncustom.com*

PEARSON
Custom Publishing

PEARSON CUSTOM PUBLISHING
501 Boylston Street, Suite 900, Boston, MA 02116
A Pearson Education Company

CONTENTS

Preface		*vii*
1	**THE DECIMAL NUMBER SYSTEM**	**1**
	Review	5
	Exercises	6
	Answers to Exercises	9
2	**ADDITION OF WHOLE NUMBERS**	**11**
	Review	14
	Exercises	15
	Answers to Exercises	18
3	**SUBTRACTION OF WHOLE NUMBERS**	**19**
	Review	23
	Exercises	24
	Answers to Exercises	27
4	**MULTIPLICATION OF WHOLE NUMBERS**	**29**
	Review	36
	Exercises	37
	Answers to Exercises	40
5	**DIVISION OF WHOLE NUMBERS**	**41**
	Review	46
	Exercises	47
	Answers to Exercises	50
6	**PRIME NUMBERS AND FACTORING**	**51**
	Review	56
	Exercises	57
	Answers to Exercises	60
7	**COMMON FRACTIONS & MIXED NUMBERS**	**61**
	Review	68
	Exercises	69
	Answers to Exercises	72
8	**THE BASIC PRINCIPLE OF FRACTIONS**	**73**
	Review	78
	Exercises	79
	Answers to Exercises	82

9	**ADDITION OF FRACTIONS**	**83**
	Review	93
	Exercises	94
	Answers to Exercises	96
10	**SUBTRACTION OF FRACTIONS**	**97**
	Review	101
	Exercises	102
	Answers to Exercises	104
11	**THE LEAST COMMON DENOMINATOR**	**105**
	Review	110
	Exercises	111
	Answers to Exercises	113
12	**MULTIPLICATION OF FRACTIONS**	**115**
	Review	122
	Exercises	123
	Answers to Exercises	124
13	**DIVISION OF FRACTIONS**	**125**
	Review	130
	Exercises	132
14	**DECIMAL FRACTIONS**	**133**
	Review	139
	Exercises	140
	Answers to Exercises	142
15	**ADDITION & SUBTRACTION OF DECIMALS**	**145**
	Review	151
	Exercises	152
	Answers to Exercises	154
16	**MULTIPLICATION OF DECIMAL FRACTIONS**	**155**
	Review	159
	Exercises	160
	Answers to Exercises	162
17	**DIVISION OF DECIMAL FRACTIONS**	**163**
	Review	168
	Exercises	169
	Answers to Exercises	171
18	**PERCENT**	**173**

Developmental Arithmetic–A Computational Review

Review	180
Exercises	181
Answers to Exercises	182

REVIEW TEST 1	**183**
REVIEW TEST 2	**185**
REVIEW TEST 3	**187**
REVIEW TEST 4	**189**
REVIEW TEST 5	**191**
REVIEW TEST 6	**193**
REVIEW TEST 7	**195**
REVIEW TEST 8	**197**
REVIEW TEST 9	**199**
REVIEW TEST 10	**201**
REVIEW TEST 11	**203**
REVIEW TEST 12	**205**
REVIEW TEST 13	**207**
REVIEW TEST 14	**209**
REVIEW TEST 15	**211**
REVIEW TEST 16	**213**
REVIEW TEST 17	**215**
REVIEW TEST 18	**217**
INDEX	**219**

preface

Developmental Arithmetic–A Computational Review 6th Edition is written for those who want to take the time to go through a brief review of the arithmetic of whole numbers, common fractions, and decimal fractions. The review begins with what is meant by a number and finishes with a review of percent.

The explanations, examples, exercises, and review tests:
- provide an opportunity for success in mathematics,
- help build a persons self-confidence, and
- open doors to further work in mathematics.

Use of the Book

Developmental Arithmetic–A Computational Review 6th Edition is designed for the first half of a semester-long review of arithmetic without the use of a calculator. A second volume, Developmental Arithmetic–Selected Applications 5th Edition, provides an opportunity to use the skills developed in the Computational Review. Together the two volumes can be used for a one semester course that prepares an individual to enter a course in pre-algebra. But the real goal of Developmental Arithmetic is to help students feel better about arithmetic by experiencing success; and as a result feel better about themselves. The hope is that the experience will then open doors that have for too long been closed.

This 6th edition includes a DVD of the lectures I gave using the text in our Introduction to Mathematics course, Math 10, at Modesto Junior College in the Fall semester of 2006. I encourage you to watch the lectures related to each lesson of the text. All you need to do is place the

DVD in a DVD ROM drive of a computer and a web page will open up. When you click on the link to each lesson you can watch a streaming video lecture using Real Network's Real One Player.

Contributors

I would like to express my heartfelt thanks to the following people who had a part in creating <u>Developmental Arithmetic–A Computational Review 6th Edition</u>:

- To the many, many students who have used the previous editions. Their often enthusiastic involvement with the text and their suggestions for improvements provided the inspiration to create a better text.

- To Darrell Top, my long time friend and now retired colleague; we together developed and taught the Introduction To Mathematics class in which we use the text. His support and encouragement were critical in making the text a reality.

- To Elzbieta Jarrett, my friend and colleague, for her ongoing participation as we continue the program in which we use the text. Her suggestions and corrections have been included in this 6th edition of the text.

- To Heather Curl Nunes, copy editor, for her critical reviews and constructive suggestions of the earlier editions of the manuscript. The results of her work are evident in the readability of the text.

- To Joshua Delahunty, illustrator and typographer, for his tireless work, commitment to excellence, and extraordinary talent in desktop publishing. His work is evident in the lay out of each lesson of the text.

lesson one

THE DECIMAL NUMBER SYSTEM

A **number** is an idea. It tells how much or how many. **Numerals** are the symbols we use to write numbers. Just as the letters of the alphabet a,b,c,d,e,... are used to write words, the digits 0,1,2,3,4,5,6,7,8,9 are used to write numerals. At the beginning of our review it is important to distinguish between a "number" and a "numeral":

Numbers Versus Numerals

- a **number** is what we think of.

- a **numeral** is what we write down.

We often use the word number when we really should say numeral. We do this because most of us feel more comfortable using the word "number". In the following lessons, we will often refer to numbers rather than making the distinction between a number and a numeral.

Our numeral system is called a **decimal numeral system** because we use ten digits (0,1,2,3,4,5,6,7,8,9) to write numbers. The decimal numeral systems based on groups of ten; so we say it is a base ten system. In the decimal numeral system, the digits are placed side by side and the position of each digit is very important.

The Decimal Numeral System

When we write a numeral such as 358, we are using what we call **positional notation;** where each position of a digit has an understood place value. Positional notation provides the advantage of allowing us to state a numeral without having to think about the value of each digit.

Positional Notation

Verbal Notation

We often write a number using **verbal notation**. Most of us use verbal notation when we write checks or use a number in conversation. Verbal notation involves stating the place value of each digit when we say the number. The three digits in the numeral 358 are written in verbal notation as:

three hundred fifty-eight.

Expanded Notation

Sometimes we write a numeral in **expanded notation** that shows us the value of each digit. We write 358 in expanded notation as follows:

$$3 \times 100 + 5 \times 10 + 8 \times 1$$

We refer to the value of each digit as the place value of the digit. For example, the place value of the 3 in 358 is 100. We find the place value of a digit using what is called a place value chart. The place value of each position is given in verbal notation in the following table.

Place Value Chart

THE PLACE VALUE CHART
billions | hundred millions | ten millions | millions | hundred thousands | ten thousands | thousands | hundreds | tens | ones

We always start with the ones place on the right side of the place value chart. As we move to the left, each new position on the chart is equal to 10 times the value of the position to the right. The place value chart is shown below using positional notation.

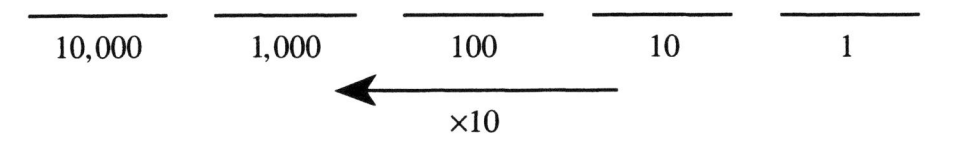

| 10,000 | 1,000 | 100 | 10 | 1 |

←―― ×10

Powers of 10

There is an easier way to write the place value chart using what we call **powers of ten**. We use an exponent (called a superscript in a word processing package) to write each place value as a power of ten. For example, 100 is written as 10^2. We read 10^2 as 10 squared or 10 to the second power. The 10 is called the base and the exponent of 2 tells us that the 10 should be written twice as a factor in the product 10×10. In the same way we can write 1,000 as 10^3 where the exponent of 3 tells us that $10^3 = 10 \times 10 \times 10$. We refer to 10^3 as 10 cubed or 10 to the third power. It should be pointed out that 10 can be written as 10^1 or 10 to the first power. We can use the powers of 10 to rewrite the place value chart.

| 10^4 | 10^3 | 10^2 | 10^1 | 1 |

Verbal Notation To Positional Notation

When we are given a numeral in verbal notation, we occasionally have problems with writing the numeral in positional notation. It is important to remember that in positional notation there must be a digit recorded for each place value position to the right of the beginning digit. For example, in positional notation the numeral for five hundred seven must show a 0 in the tens position. The beginning digit is the 5 in the hundreds position; there are 0 tens and 7 ones. So in positional notation, the numeral for five hundred seven is 507. When we write the numeral 507 in expanded notation, it is easier to see what is involved.

$$5 \times 100 + 0 \times 10 + 7 \times 1$$

Lesson One – The Decimal Number System

Positional Notation To Verbal Notation

A different type of problem faces us when we must change from positional notation to verbal notation. Commas are used to help us change from positional notation to verbal notation. We need to remember that commas are placed before every three digits going from the right to the left. For example, the numeral 28762684 is written as

28,762,684

The key to writing the numeral in verbal notation is to work with each group of three digits to the left of the commas. We always start with the far left side of the numeral. We look at the value of the first digit to the left of the first comma on the left.

In the numeral 28,762,684 the 8 is the first digit to the left of the first comma on the left. The 8 is in the millions position on the place value chart. So we start by saying twenty-eight million.

The next group of three digits is 762. The 2 is the digit in front of the comma. We see that the 2 is in the thousands position on the place value chart. So we continue; saying seven hundred sixty-two thousand.

The last group of three digits at the far right side of the numeral reads as six hundred eighty four. The numeral 28,762,684 is read in verbal notation as follows:

twenty-eight million, seven hundred sixty-two thousand, six hundred eighty-four.

Please note that we did not use the word "and" as we wrote the numeral in verbal notation. We will wait to use "and" in verbal notation to represent a decimal in a later lesson when we work with decimal fractions.

REVIEW

- **Positional notation:** The way we actually write a numeral when we use the digits between 0 and 9, for example 358.   *See page 1*

- **Place value**
  The value of each digit in a number is determined by the position of the digit in the place value chart; for example the place value of the 5 in 358 is 10. We say that 5 is in the 10's place or that 10 is the place value of the 5.   *See page 2*

- **Expanded notation:** Shows the place value of each digit of a number, for example $358 = 3 \times 100 + 5 \times 10 + 8 \times 1$ *See page 2*

- **Verbal notation:** Used to write a number as it would be said giving the place value of each digit, for example 358 is read as three hundred fifty-eight.   *See page 2*

- **Powers of 10**
  We write a 1 followed by zeros using the base 10 and an exponent equal to the number of zeros after the one, for example $10,000 = 10^4$.   *See page 3*

Lesson One – The Decimal Number System

EXERCISES

1. Write the following in positional notation:

   a. Two thousand forty-five

   2,045

   b. Fourteen thousand three

   14,003

   c. One million seventy-three thousand four hundred two

   1,073,402

   d. Fifteen billion two million four thousand twenty-nine

   15,002,004,029

2. Write the following numbers in expanded notation:

   a. 3,476

   $3 \times 1000 + 4 \times 100 + 7 \times 10 + 6 \times 1$

   b. 7,823,491

   $7 \times 1,000,000 + 8 \times 100,000 + 2 \times 10,000 + 3 \times 1000 + 4 \times 100 + 9 \times 10 + 1 \times 1$

   c. 706,503

   $7 \times 100,000 + 6 \times 1000 + 5 \times 100 + 3 \times 1$

   d. 67,500,327

   $6 \times 10,000,000 + 7 \times 1,000,000 + 5 \times 100,000 + 3 \times 100 + 2 \times 10 + 7 \times 1$

Developmental Arithmetic—A Computational Review

3. Write the following numerals in verbal notation:

   a. 6,039

   six thousand thirty nine

   b. 1,652,304

   one million six hundred fifty two thousand three hundred four

   c. 354,006,007

   three hundred fifty four million six thousand seven

   d. 1,072,034,086

   one billion seventy two million thirty four thousand eighty six

4. In the numeral 74,509 use the positional chart or expanded notation to find the place value of:

   a. the 5

   hundreds

   b. the 7

   ten thousands

   c. the 9

   ones

   d. the 0

   tenths

Lesson One – The Decimal Number System

5. Write each of the following as a power of 10:

   a. 1,000

   10^3

   b. 10,000,000

   10^7

   c. 100,000,000,000

   10^{11}

   d. 1,000,000,000,000,000

   10^{15}

6. Insert commas in the following numerals and write the numeral in verbal notation:

   a. 4 3 5 2

   4,352

   b. 1 0 7 0 0 3

   107,003

   c. 5 0 9 7 6 0 5

   5,097,605

   d. 1 3 0 7 8 6 0 4 5 2

   1,307,860,452

ANSWERS TO EXERCISES

1. a. 2,045   b. 14,003   c. 1,073,402   d. 15,002,004,029

2. a. $3 \times 10^3 + 4 \times 10^2 + 7 \times 10^1 + 6 \times 1$

   b. $7 \times 10^6 + 8 \times 10^5 + 2 \times 10^4 + 3 \times 10^3 + 4 \times 10^2 + 9 \times 10^1 + 1 \times 1$

   c. $7 \times 10^5 + 0 \times 10^4 + 6 \times 10^3 + 5 \times 10^2 + 0 \times 10^1 + 3 \times 1$

   d. $6 \times 10^7 + 7 \times 10^6 + 5 \times 10^5 + 0 \times 10^4 + 0 \times 10^3 + 3 \times 10^2 + 2 \times 10^1 + 7 \times 1$

3. a. six thousand thirty-nine

   b. one million six hundred fifty-two thousand three hundred four

   c. three hundred fifty-four million six thousand seven

   d. one billion seventy-two million thirty-four thousand eighty-six

4. a. hundreds   b. ten thousands   c. ones   d. tens

5. a. 10^3 b. 10^7 c. 10^{11} d. 10^{15}

6. a. 4,352
   four thousand three hundred fifty-two

   b. 107,003
   one hundred seven thousand three

   c. 5,097,605
   five million ninety-seven thousand six hundred five

   d. 1,307,860,452
   one billion three hundred seven million eight hundred sixty thousand four hundred fifty-two

Lesson One – The Decimal Number System

lesson two

ADDITION OF WHOLE NUMBERS

Addition is used to find the **sum** of two or more numbers. The sum is the total of the numbers added together. In order to add whole numbers, we must be careful to line up the numbers correctly. As an example, we will consider the following addition problem:

$$153 + 24 + 8,679$$

The numbers need to be rewritten in a vertical arrangement so that each digit is in the correct position for its place value. We need to remember that the last digit to the right of each number is in the ones place; so we line up the digits in the ones place.

$$\begin{array}{r} 153 \\ 24 \\ \underline{8679} \end{array}$$

Each column can now be added. The columns are added from either top to bottom or bottom to top. We begin at the far right with the ones column and add:

$$3 + 4 + 9 = 16$$

Remember the 16 is 1 ten and 6 ones. We write the 6 under the ones column and carry the 1 ten to the tens column as shown below.

Sum

Line Up The Ones Column To Add Whole Numbers.

Lesson Two—Addition of Whole Numbers 11

$$\begin{array}{r} \overset{1}{1\,5\,3} \\ 2\,4 \\ 8\,6\,7\,9 \\ \hline 6 \end{array}$$

Always Carry To The Column To The Left

We then add the digits in the tens column; be sure to include the 1 we **carried** from the sum of the ones column.

$$1 + 5 + 2 + 7 = 15$$

This time, we must remember that the 15 means 1 hundred and 5 tens; so the 1 is carried to the column in the hundreds place as shown:

$$\begin{array}{r} \overset{1\;1}{1\,5\,3} \\ 2\,4 \\ 8\,6\,7\,9 \\ \hline 5\,6 \end{array}$$

We continue to add the columns and find the sum of 8,856:

$$\begin{array}{r} \overset{1\;1}{1\,5\,3} \\ 2\,4 \\ 8\,6\,7\,9 \\ \hline 8\,8\,5\,6 \end{array}$$

Reverse The Addition To Check Your Answer

When we do addition, it is easy to make mistakes as we find the sum for each column. As such, it is important to check our work. If we find sums by adding from the top down, we can check our work by adding the columns from the bottom to the top.

Of course, it would help even more if we had a way of finding the sum for each column that would reduce the chance of making an error. There are two methods that we can use. The first one is called the **low stress algorithm for addition**; it involves writing down sub-totals as we add the digits in each column. The following problem shows how the low stress algorithm works. We will find the sums going from the top to the bottom for each column.

The Low Stress Algorithm

```
         2
       3   4
    5  5   2
   10        6
   14  4   6  12
   21  7   8  20
   24  3   5  25
      ─────────
       2  4  5
```

The Low Stress Algorithm

Another method that we can use to add colums of numbers involves **grouping by tens**. Again, this is a method that can help us add more quickly and more accurately. Grouping by tens means that we find numbers that add up to ten in each column. We then add the tens with whatever is left in the column. Study the example below:

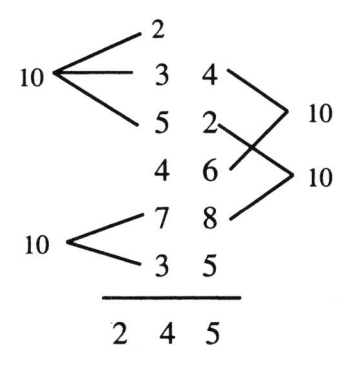

```
              2
   10 ⟨  3   4
          5   2 ⟩ 10
              4   6 ⟩ 10
         ⟨ 7   8
   10 ⟨  3   5
         ─────────
          2  4  5
```

The Grouping By Tens Method

Lesson Two—Addition of Whole Numbers

REVIEW

See Page 11
- **Sum**
  The word sum is used for the answer to an addition problem.

See Page 11
- **Addition of Whole Numbers**
  Always line up numbers so that all the digits in a column have the same place value. This is done by placing the ones digit at the far right of each number in the same column.

See Page 12
- **Carries in Addition**
  Begin addition in the ones column. Place the carries from each sum at the top of the next column to the left.

See Page 12
- **Checking Answers in Addition**
  Check answers by reversing the order of addition.

See Page 12
- **The Low Stress Algorithm for Addition**
  Use either the low stress algorithm or grouping by tens to add the columns quickly and accurately.

EXERCISES

1. Find the following sums. Try to find the thirty answers in less than one minute with out any mistakes.

$$\begin{array}{r}4\\+\,0\\\hline\end{array}$$ 4
$$\begin{array}{r}8\\+\,5\\\hline\end{array}$$ 13
$$\begin{array}{r}2\\+\,9\\\hline\end{array}$$ 11
$$\begin{array}{r}5\\+\,4\\\hline\end{array}$$ 9
$$\begin{array}{r}1\\+\,6\\\hline\end{array}$$ 7

$$\begin{array}{r}7\\+\,9\\\hline\end{array}$$ 16
$$\begin{array}{r}6\\+\,4\\\hline\end{array}$$ 10
$$\begin{array}{r}8\\+\,0\\\hline\end{array}$$ 8
$$\begin{array}{r}7\\+\,7\\\hline\end{array}$$ 14
$$\begin{array}{r}9\\+\,5\\\hline\end{array}$$ 14

$$\begin{array}{r}9\\+\,9\\\hline\end{array}$$ 18
$$\begin{array}{r}6\\+\,5\\\hline\end{array}$$ 11
$$\begin{array}{r}2\\+\,8\\\hline\end{array}$$ 10
$$\begin{array}{r}4\\+\,7\\\hline\end{array}$$ 11
$$\begin{array}{r}9\\+\,6\\\hline\end{array}$$ 15

$$\begin{array}{r}3\\+\,7\\\hline\end{array}$$ 10
$$\begin{array}{r}9\\+\,4\\\hline\end{array}$$ 13
$$\begin{array}{r}3\\+\,3\\\hline\end{array}$$ 6
$$\begin{array}{r}8\\+\,8\\\hline\end{array}$$ 16
$$\begin{array}{r}4\\+\,1\\\hline\end{array}$$ 5

$$\begin{array}{r}8\\+\,7\\\hline\end{array}$$ 15
$$\begin{array}{r}3\\+\,9\\\hline\end{array}$$ 12
$$\begin{array}{r}5\\+\,3\\\hline\end{array}$$ 8
$$\begin{array}{r}8\\+\,9\\\hline\end{array}$$ 17
$$\begin{array}{r}6\\+\,7\\\hline\end{array}$$ 13

$$\begin{array}{r}8\\+\,4\\\hline\end{array}$$ 12
$$\begin{array}{r}7\\+\,9\\\hline\end{array}$$ 16
$$\begin{array}{r}8\\+\,3\\\hline\end{array}$$ 11
$$\begin{array}{r}5\\+\,7\\\hline\end{array}$$ 12
$$\begin{array}{r}3\\+\,0\\\hline\end{array}$$ 3

Work For Accuracy And Quickness

Lesson Two—Addition of Whole Numbers

2. Use the low stress algorithm to add the following numbers from the top down. Check your answer for each column by adding from the bottom to the top.

4	5	9	2
0	7	4	7
7	3	8	0
6	9	7	9
4	8	3	3
+ 3	+ 6	+ 5	+ 8
24	38	37	29

3. Find the sums; first rewrite each problem lining up the columns.

a. $23 + 187 + 9 + 3{,}026 + 87$

```
   23
  187
    9
 3026
+  87
─────
 3332
```

b. $10{,}603 + 5{,}679 + 246 + 17 + 8{,}931$

```
 10,603
  5,679
    246
     17
  8,931
 ──────
 25476
```

c. $9{,}703 + 398 + 7{,}876 + 69 + 15{,}452$

```
  9,703
    398
  7,876
     69
 15,452
 ──────
  33498
```

4. Use grouping by tens to find the sums for each problem.

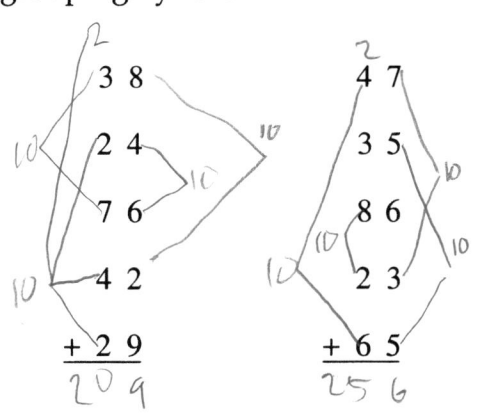

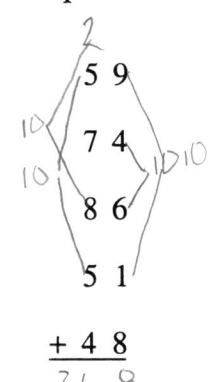

```
   3 8         4 7         5 9
   2 4         3 5         7 4
   7 6         8 6         8 6
   4 2         2 3         5 1
 + 2 9       + 6 5       + 4 8
  209         256         318
```

5. Add; show carry for each column.

a.
```
  1 2
  5 2 6
    1 5
  8 7 8
+   5 7
 1 4 7 6
```

b.
```
  2 2 7
  1,7 0 4
      4 5 6
  2 8,9 7 2
+     1,0 9 3
  3 2,2 2 5
```

c.
```
  2 2 1 1
  8,6 0 2
      5 8 7
  3 7,9 6 4
+ 1 5,0 4 3
  6 2,1 9 6
```

6. An interesting sequence of numbers was first given by Leonardo de Pisa in 1206 in a book called Liber Abaci. The first five numbers are 1, 1, 2, 3, 5. The next number is found by adding 3 + 5 = 8. Each new number is found by adding the two previous numbers. This sequence is known as the Fibonacci Sequence. It was given by Leonardo (known as Fibonacci) as the solution to a problem about how many rabbits one would have at the end of one year if one started with one pair of rabbits. Find the next ten numbers in the sequence.

1 1 2 3 5 8 13 21 34 55 89 144 233 377 610 987

Lesson Two—Addition of Whole Numbers

ANSWERS TO EXERCISES

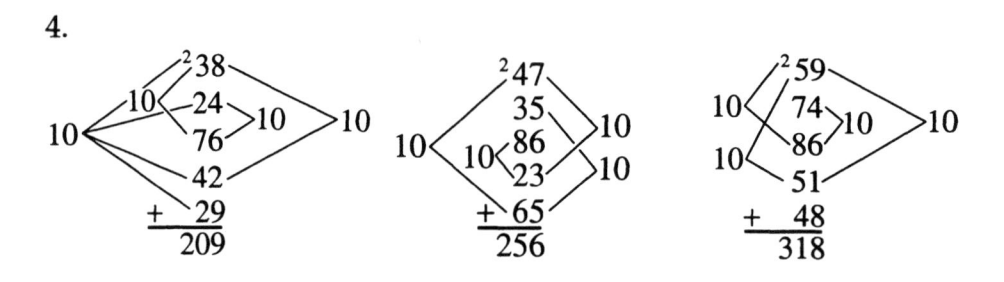

1.

4	13	11	9	7
16	10	8	14	14
18	11	10	11	15
10	13	6	16	5
15	12	8	17	13
12	16	11	12	3

2. 24   38   36   29

3. a.
```
    23
   187
     9
 3,026
+   87
─────
 3,332
```

b.
```
10,603
 5,679
   246
    17
+8,931
──────
25,476
```

c.
```
 9,703
   398
 7,876
    69
+15,452
───────
 33,498
```

4. [diagrams showing addition with carrying: first column sums to 209 with addends 38, 24, 76, 42, 29; second column sums to 256 with addends 47, 35, 86, 23, 65; third column sums to 318 with addends 59, 74, 86, 51, 48]

5. a. 1,476   b. 32,225   c. 62,196

6. 13, 21, 34, 55, 89, 144, 233, 377, 610, 987

lesson three

SUBTRACTION OF WHOLE NUMBERS

Subtraction is a comparison of two numbers. The smaller number is subtracted from the larger number to find the **difference** between the two. In order to subtract, it is necessary to know the addition facts. Actually, every addition fact has two related subtraction facts. For example, knowing that $3 + 4 = 7$ tells us that $7 - 4 = 3$ and $7 - 3 = 4$. Because of this relationship between addition and subtraction, subtraction is called the **inverse** of addition.

Difference

Inverse

Once again it is very important that the numerals in a subtraction problem are lined up so that each digit is in the correct position for its place value. We then subtract the digits in each column, starting at the far right or ones column. For example $1,859 - 235$ is done as follows:

$$\begin{array}{r} 1,859 \\ -235 \\ \hline 1,624 \end{array}$$

In subtraction it may be necessary to **borrow** from the column with higher place value in order to complete the subtraction. To understand borrowing, we really need an understanding of the place value of each position. Let's try an example.

Borrowing

Subtract 164 from 7,352

Line up the numerals.
$$\begin{array}{r} 7,352 \\ -164 \\ \hline \end{array}$$

Lesson Three—Subtraction of Whole Numbers

Begin at the right (the ones column). Since it is not possible to subtract 4 from 2, it is necessary to borrow one 10 from the top digit in the next column. Since we are borrowing a 10, we have 10 + 2 or 12. Now we can subtract 4 from 12 and place the 8 under the ones column.

$$\begin{array}{r} 73\overset{4}{\cancel{5}}{}^{1}2 \\ -\ 1\ 6\ 4 \\ \hline 8 \end{array}$$

In the tens column, we are not able to subtract 6 from 4; again it is necessary to borrow. We will borrow one 100 from the hundreds column, we find that one hundred is 10 tens; we now have 10 + 4 or 14 tens. Subtracting 6 from 14 gives us 8 in the tens column.

$$\begin{array}{r} \overset{\ \ \ 1}{\underset{2}{\cancel{\ }}}4 \\ 7\,\overset{}{\cancel{3}}\,\overset{}{\cancel{5}}\,{}^{1}2 \\ -\ 1\ 6\ 4 \\ \hline 8\ 8 \end{array}$$

Completing the problem 2 − 1 = 1 and 7 − 0 = 7.

$$\begin{array}{r} \overset{\ \ \ 1}{\underset{2}{\cancel{\ }}}4 \\ 7\,\overset{}{\cancel{3}}\,\overset{}{\cancel{5}}\,{}^{1}2 \\ -\ 1\ 6\ 4 \\ \hline 7{,}1\ 8\ 8 \end{array}$$

Working With Zeros

Sometimes zeros in a subtraction problem are a source of anxiety. Some zeros are easily dealt with as in the example below.

$$\begin{array}{r} \overset{3}{\cancel{4}}{}^{1}0\ 3 \\ -\ 1\ 6\ 2 \\ \hline 2\ 4\ 1 \end{array}$$

We simply borrow 1 from the 4 so that 6 can be subtracted from 10. However, in the problem below, more steps are required. We first line up the digits with the larger numeral on top and the smaller numeral on the bottom.

$$\begin{array}{r} 6\,0\,3 \\ -\ 2\,8\,7 \\ \hline \end{array}$$

Because it is necessary to borrow before we can subtract in the ones column, we look to the tens column. But it is a zero so we must go to the hundreds column to borrow from the 6 in the hundreds column.

$$\begin{array}{r} \overset{5}{\cancel{6}}{}^{1}0\,3 \\ -\ 2\,8\,7 \\ \hline \end{array}$$

Now it is possible to borrow 1 ten from the 10 tens in the tens column, leaving 9 in the tens column and 13 in the ones column.

$$\begin{array}{r} \overset{5}{\cancel{6}}\ {}^{9}\!{}^{1}\!\cancel{0}\,{}^{1}3 \\ -\ 2\,8\,7 \\ \hline 3\,1\,6 \end{array}$$

If a subtraction problem has more than one zero in its top numeral, the same procedure is used but it involves more borrowing steps. Study the following example:

$$\begin{array}{r} 6\,0\,0\,3 \\ -\ 3\,2\,5\,6 \\ \hline \end{array} \quad \begin{array}{r} \overset{5}{\cancel{6}}{}^{1}0\,0\,3 \\ -\ 3\,2\,5\,6 \\ \hline \end{array} \quad \begin{array}{r} \overset{5}{\cancel{6}}\,{}^{9}\!\cancel{0}\,{}^{1}0\,3 \\ -\ 3\,2\,5\,6 \\ \hline \end{array} \quad \begin{array}{r} \overset{5}{\cancel{6}}\,{}^{9}\!\cancel{0}\,{}^{9}\!\cancel{0}\,{}^{1}3 \\ -\ 3\,2\,5\,6 \\ \hline 2\,7\,4\,7 \end{array}$$

Subtraction can be checked by adding the difference (answer) to the smaller number (the bottom numeral). If the result of the addition is the larger number (the top numeral), then the subtraction is correct. This is illustrated in the following example.

Checking Subtractions

$$\begin{array}{r} \text{Subtract}\ \begin{array}{r} 217 \\ -\ 103 \\ \hline 114 \end{array} \rightarrow \begin{array}{r} 114 \\ \text{Add}\ +\ 103 \\ \hline 217 \end{array} \end{array} \bigg)\ \text{Answers Check}$$

Lesson Three—Subtraction of Whole Numbers

The Austrian Method

The **Austrian Method** of subtraction is based on the idea of addition rather than subtraction. This method avoids borrowing and instead we carry as we do in addition. Let's try the example

$$\begin{array}{r} 23 \\ -7 \\ \hline \end{array}$$

Using the Austrian method we ask, "What number added to 7 will give us a number that ends in 3?" Since 6 + 7 = 13, we write 6 under the ones column and carry the 1 to the tens column as shown.

$$\begin{array}{r} 23 \\ -^17 \\ \hline 6 \end{array}$$

Now the question is, "What is added to 1 to get 2?" The answer, of course, is 1 and a 1 is placed under the tens column for a difference of 16.

$$\begin{array}{r} 23 \\ -^17 \\ \hline 16 \end{array}$$

Study the examples below to help you understand the Austrian Method more clearly.

$$\begin{array}{r} 745 \\ {}^{4\,9} \\ -\,386 \\ \hline 359 \end{array} \qquad \begin{array}{r} 6001 \\ {}^{4\,3\,6} \\ -\,3256 \\ \hline 2745 \end{array}$$

The Austrian Method has the advantages of:

1. Using the addition facts you already know.
2. Not requiring borrowing.

If subtraction causes a sense of anxiety in you, maybe the Austrian Method is for you. Try it!

REVIEW

- **Subtraction of Whole Numbers**
  Always line up numerals so that all the digits in a column have the same place value. This is done by placing the ones digit at the far right of each number in the same column. The larger numeral is placed on the top of the smaller numeral.

  See page 19

- **Difference**
  The word difference is used for the answer to a subtraction problem.

  See page 19

- **Borrowing**
  Begin subtraction in the ones column. Borrowing is necessary if the bottom digit in a column is larger than the top digit.

  See page 19

- **Checking Answers in Subtraction**
  Check answers by adding the answer to the number that was subtracted. The result should be the top number.

  See page 21

- **Austrian Method**
  The Austrian Method of subtraction is an alternative way of doing subtraction using addition instead of borrowing.

  See page 22

EXERCISES

1. Do the following subtraction problems as quickly and as accurately as possible.
   (You might want to try them using the Austrian Method.)

Work For Accuracy And Quickness

13 − 7 = 6	8 − 0 = 8	15 − 6 = 9	16 − 9 = 7	7 − 7 = 0
11 − 5 = 6	18 − 9 = 9	10 − 3 = 7	12 − 4 = 8	18 − 8 = 10
9 − 6 = 3	17 − 8 = 9	13 − 8 = 5	11 − 2 = 9	14 − 7 = 7
12 − 9 = 3	9 − 7 = 2	14 − 9 = 5	3 − 0 = 3	13 − 5 = 8
15 − 9 = 6	9 − 1 = 8	19 − 9 = 10	16 − 8 = 8	10 − 4 = 6
17 − 9 = 8	13 − 6 = 7	5 − 0 = 5	11 − 6 = 5	12 − 7 = 5

2. Subtract:

a. 39
   − 24
   ‾‾‾
   15

b. 85
   − 30
   ‾‾‾
   55

c. 19
   − 9
   ‾‾
   10

d. ⁴⁄5̶7
   − 39
   ‾‾‾
   18

e. ⁷⁄8̶0
   − 56
   ‾‾‾
   24

3. Subtract:

a. 5⁶7̶6
   − 208
   ‾‾‾‾
   368

b. ²⁄³⁰̶⁹4
   − 156
   ‾‾‾‾
   148

c. ⁶⁄7̶,⁹⁄0̶⁰⁄0̶3
   − 3,257
   ‾‾‾‾‾‾
   3746

d. ⁵⁄6̶0,³⁄4̶05
   − 43,020
   ‾‾‾‾‾‾‾
   17385

4. Subtract and check:

a. ⁶⁄8̶⁵⁄7̶,657
   − 64,986
   ‾‾‾‾‾‾‾
   22671

b. ⁵⁄6̶⁵,⁴⁄4̶¹³⁄8̶9
   − 47,899
   ‾‾‾‾‾‾‾
   17590

c. ⁸⁄9̶⁹⁄0̶⁶⁄,5̶⁴⁄7̶⁶⁄2̶
   − 658,394
   ‾‾‾‾‾‾‾‾
   248178

d. ⁰⁄1̶,²⁄3̶⁹⁄0̶⁹⁄0̶³⁄,4̶⁹⁄0̶5
   − 763,876
   ‾‾‾‾‾‾‾‾‾
   536529

If you use the Austrian Method you do not need to check

Lesson Three—Subtraction of Whole Numbers

Remember to add, not borrow

5. Subtract using the Austrian Method. (Do not borrow.)

            4,302   5,002   8,010,203
a. − 1,654   b. − 3,674   c. − 4,675,089

          2648   1328   3335114

6. Rewrite each problem lining up the columns:

a. Find the difference of 10,576,324 and 7,038,517.

    10,576,324
  − 7,038,517
     3,537,807

b. Subtract 8,763 from 107,002.

    107,002
  −   8,763
      98,339

Developmental Arithmetic–A Computational Review

ANSWERS TO EXERCISES

1.

6	8	9	7	0
6	9	7	8	10
3	9	5	9	7
3	2	5	3	8
6	8	10	8	6
8	7	5	5	5

2. a. 15   b. 55   c. 10   d. 18   e. 24

3. a. 368   b. 148   c. 3,746   d. 17,385

4. a. 22,671   b. 17,590   c. 248,178   d. 536,529

5.

a. 4302
    2 7 6
    − 16̸5̸4
    ─────
    2648

b. 5002
    4 7 8
    − 3̸6̸74
    ─────
    1328

c. 8 0 1 0 2 0 3
    5 7 8 1 9
    − 4̸6̸7̸5̸0̸8̸9
    ─────────
    3 3 3 5 1 1 4

6.

a. 10,576,324
    − 7,038,517
    ──────────
    3,537,807

b. 107,002
    − 8,763
    ─────────
    98,239

Lesson Three—Subtraction of Whole Numbers

lesson four

MULTIPLICATION OF WHOLE NUMBERS

Numbers that are multiplied together are called **factors**. When we multiply the factors, we get their **product**. For example, in the multiplication problem $3 \times 7 = 21$, 3 and 7 are factors and 21 is their product.

Factors

The Answer To A Multiplication Is Called A Product

Multiplication is a shortcut to addition. The problem 3×7 can be thought of as addition, where the three 7s are added as follows:

$$3 \times 7 = 7 + 7 + 7 = 21$$

Instead of adding, we memorize the multiplication fact $3 \times 7 = 21$.

It is important to understand that **reversing the order of the factors in a multiplication problem** does not change the product. We can see this when we find 7×3.

Multiplications Can Be Reversed

$$7 \times 3 = 3 + 3 + 3 + 3 + 3 + 3 + 3 = 21$$

$$3 \times 7 = 7 \times 3$$

Although products can be found by adding, it is a real time saver to memorize the products of the numbers between 0 and 10. There is a multiplication table on the next page. If you have forgotten some of the products or never did take time to learn them before, now is the time to do so. You really do need to spend the necessary time memorizing the multiplication facts. Sometimes an old-fashioned idea like creating a set of flash cards out of 3×5 cards can help. Maybe it is just taking the time

Lesson Four—Multiplication of Whole Numbers

to focus on a few multiplication facts that you have forgotten. Whatever the case, please take the time.

Do memorize the multiplication facts from 0 to 10.

Maybe a quick review of the multiplication table will jar your memory. Remember that reversing the order of the factors does not change the product, so you only need to memorize half of the table. For example both 7×8 and 8×7 give the same product of 56. So you only need to memorize half of the table.

MEMORIZE THE TABLE

x	0	1	2	3	4	5	6	7	8	9	10
0	0	0	0	0	0	0	0	0	0	0	0
1	0	1	2	3	4	5	6	7	8	9	10
2	0	2	4	6	8	10	12	14	16	18	20
3	0	3	6	9	12	15	18	21	24	27	30
4	0	4	8	12	16	20	24	28	32	36	40
5	0	5	10	15	20	25	30	35	40	45	50
6	0	6	12	18	24	30	36	42	48	54	60
7	0	7	14	21	28	35	42	49	56	63	70
8	0	8	16	24	32	40	48	56	64	72	80
9	0	9	18	27	36	45	54	63	72	81	90
10	0	10	20	30	40	50	60	70	80	90	100

The products on the diagonal

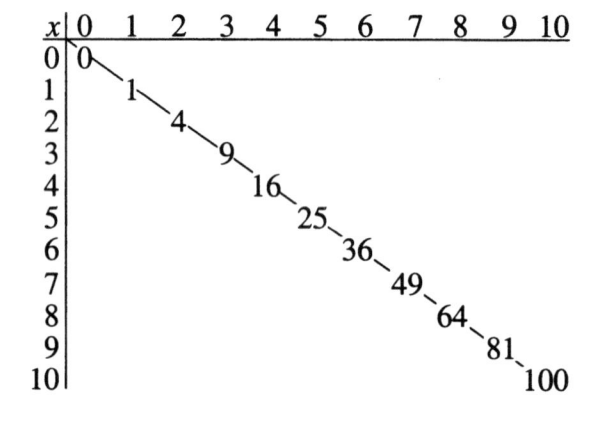

Perfect Squares

x	0	1	2	3	4	5	6	7	8	9	10
0	0										
1		1									
2			4								
3				9							
4					16						
5						25					
6							36				
7								49			
8									64		
9										81	
10											100

are particularly important. These numbers are called **perfect squares**.

Developmental Arithmetic–A Computational Review

Multiplying a number by itself is called **squaring the number**. For example 3×3 is called 3 squared. We again use an exponent to write a perfect square.

Squaring Numbers

$$3 \times 3 = 3^2 = 9$$

Our answer 9 is called a **perfect square**. If you take the time to first memorize the perfect squares, the results can be used to help get some of the other products that may be troublesome.

Perfect Squares

For example, if there is a problem with remembering the answer for 7×8, we can use 7^2 to get our answer for 7×8. We just add one more 7 to $7 \times 7 = 7^2 = 49$.

$$7 \times 8 = 7^2 + 7 = 49 + 7 = 56$$

It may be easier to add $49 + 7$ to get the 56 then to memorize both the multiplication facts 7×7 and 7×8.

There is one more shortcut that you might want to try. It involves the multiplication facts that have 9 as a factor. The following example shows how the shortcut works. We want the answer for the product 9×8. We find the answer by first writing $8 - 1 = 7$. We then note that $7 + 2 = 9$. The first subtraction gave us the first digit of our answer 72. The second digit of our answer was the number we added to 7 to get 9. The next few examples illustrate the shortcut.

The Short Cut For 9s

$$9 \times 6$$
$$6 - 1 = ⑤ \qquad 5 + ④ = 9$$
$$54$$

$$9 \times 7$$
$$7 - 1 = ⑥ \qquad 6 + ③ = 9$$
$$63$$

Lesson Four—Multiplication of Whole Numbers

The Multiplication Algorithm

When multiplying larger numbers together, the problems could be done by adding. But the problems would require too much time to get the answer. Imagine multiplying 23 × 57 and having to write 57 twenty-three times and then adding. Or worse yet writing 23 fifty-seven times before adding. Fortunately there is an algorithm we can use to find the product of such numbers. We use the following rules:

1. Multiply the top number by each digit of the bottom number, beginning with the digit in the ones position. The first product (called a partial product) is written so that the first digit we get is in the ones position.
2. The next partial product is written so that the first digit we get is in the tens position.
3. Continue indenting one position to the left with each new partial product.
4. Add the rows of partial products to get the answer.

Follow along with the example below to see how the multiplication algorithm is used to find the products for large factors.

$$\begin{array}{r} 57 \\ \times\ 23 \end{array}$$

3 × 7 = 21 so the 1 is placed under the ones column and the 2 is carried to the tens column.

$$\begin{array}{r} {}^{2} \\ 57 \\ \times\ 23 \\ \hline 1 \end{array}$$

3 × 5 = 15 plus the 2 carried from the preceding step is 17.

```
  1 2
   57
×  23
  ───
  171
```

$2 \times 7 = 14$ The second row of multiplication is indented one digit to the left, so the 4 is placed under the 7 in the tens column and 1 is carried.

```
    1
  1 2
   57
×  23
  ───
  171
    4
```

$2 \times 5 = 10$ plus 1 carried is 11.

```
  1 1
  1 2
   57
×  23
  ───
  171
  114
```

Add the rows of partial products for the product of $57 \times 23 = 1311$.

```
    1
  1 1
  1 2
   57
×  23
  ───
  171
  114
  ────
  1311
```

Lesson Four–Multiplication of Whole Numbers

Two more examples show what to do with zeros in multiplication.

$$\begin{array}{r} \overset{\overset{4}{}}{\overset{1\,1}{308}} \\ \times 52 \\ \hline 616 \\ 15\,40 \\ \hline 16{,}016 \end{array} \qquad \begin{array}{r} \overset{\overset{1}{}\overset{2}{}}{\overset{42\,5}{1{,}527}} \\ \times 380 \\ \hline 0\,000 \\ 122\,16 \\ 458\,1 \\ \hline 580{,}260 \end{array}$$

Multiplying By Powers of Ten

There are several shortcuts that can be used when multiplying. One shortcut is used when multiplying any whole number by 10 or 100 or 1,000 or any other power of 10. We simply add zeros to the end of the whole number. For example, if we multiply $24 \times 1{,}000$, the product is found by writing the same number of zeros after 24 as follows the 1 in 1,000. The product is 24,000. If we write the original problem as 24×10^3 it is even easier to use the shortcut. The exponent of 3 tells us to add three zeros to the 24 to get 24,000. Of course we can always do the multiplication using the algorithm as follows:

$$\begin{array}{r} 1{,}000 \\ \times 24 \\ \hline 4\,000 \\ 20\,00 \\ \hline 24{,}000 \end{array}$$

Multiplying By A Number With Zeros On The End Of The Number

Another shortcut you should find worthwhile involves multiplying any whole number by a number with zeros on the end of the number. In the earlier example $1{,}527 \times 380$ instead of beginning the multiplication with the zero in the ones position, we will rewrite the problem as shown below, shifting the zero to the right.

$$\begin{array}{r} \overset{12}{\underset{1425}{1527}} \\ \times 380 \\ \hline 122\,16 \\ 458\,1 \\ \hline 580,260 \end{array}$$

A second example shows how we do the problem if there are more zeros at the end of one of the numbers.

$$\begin{array}{r} \overset{1}{\underset{321}{3742}} \\ \times 215\,000 \\ \hline 18710 \\ 3742 \\ 7484 \\ \hline 804,530,000 \end{array}$$

There is one more topic to be covered. It involves multiplying three numbers together such as $2 \times 3 \times 5$. Factors can be multiplied in any order. We show this using parentheses as follows:

Multiplying Three Numbers Together

$$(2 \times 3) \times 5 = \qquad 2 \times (3 \times 5) =$$
$$6 \times 5 = \qquad\qquad 2 \times 15 =$$
$$30 \qquad\qquad\qquad 30$$

Actually it would be better to rewrite the multiplication so that the 2 and the 5 are multiplied first as follows:

$$3 \times (2 \times 5) =$$
$$3 \times 10 =$$
$$30$$

If we have to multiply more than three numbers together, we again can multiply in any order we want. The key is to arrange the factors in a way that makes doing the multiplication as easy as possible.

Multiplying More Than Three Numbers Together

Lesson Four–Multiplication of Whole Numbers

REVIEW

See page 29

- **Factors**

  The numbers we multiply together are called factors. The factors can be multiplied together in any order.

See page 29

- **Product**

  The word product is used for the answer to a multiplication problem.

See page 30

- **Perfect Squares**

  Multiplying a number by itself is called squaring a number. The product is called a perfect square.

See page 32

- **Multiplication Algorithm for Whole Numbers**

  Multiplying large numbers requires that a top number be multiplied by each digit of the bottom number. Indenting each of those partial products one position to the left. Adding the partial products will give the product of the two original factors.

See page 34

- **Shortcuts Involving Zeros**

  Shortcuts for numbers that end in zeros can be valuable time savers. The shortcut using powers of ten requires that we add zeros to the end of one of the numbers.

EXERCISES

1. Find the following products as quickly and accurately as possible.

9 × 7 63	6 × 8 48	4 × 5 20	9 × 1 9	9 × 8 72
7 × 0 0	7 × 6 42	9 × 9 81	5 × 5 25	8 × 4 32
7 × 8 56	8 × 8 64	6 × 9 54	7 × 4 28	9 × 5 45
7 × 7 49	5 × 6 30	8 × 7 56	3 × 6 18	1 × 8 8
8 × 5 40	7 × 3 21	4 × 9 36	0 × 5 0	8 × 3 24
7 × 1 7	6 × 6 36	3 × 9 27	5 × 7 35	0 × 9 0

Work For Accuracy And Quickness

Lesson Four–Multiplication of Whole Numbers 37

2. Multiply:

```
    86
  × 57
   602
   430
  4902
```

```
    79
  × 65
   395
   474
  5135
```

```
    29
  × 76
   174
   203
  2204
```

```
    85
  × 39
   765
   255
  3315
```

3. Multiply:

```
    496
  × 819
   4964
    496
   3968
  406224
```

```
    305
  ×  78
   2400
   2139
  23790
```

```
    804
  ×  35
   4020
   2412
  28140
```

```
    401
  × 937
   2807
   1203
   3609
  375737
```

4. Find the product:

    a. 125×100

    <the handwritten work shows>
    125
    × 100
    ─────
    000
    000
    125
    ─────
    12500

    b. $703 \times 100,000$

    100000
    × 703
    ─────
    300000
    000000
    700000
    ─────
    70300000

    c. 37×10^3

    37000

    d. 914×10^6

    914000000

5. Multiply:

    a. 1,259
       × 250
       ──────
         0000
         6295
         2518
       ──────
        314750

    b. 73,679
       × 3,900
       ───────
       73679
       × 3900
       ───────
        00000
        00000
        663111
        221037
       ─────────
        287348100

Lesson Four–Multiplication of Whole Numbers

ANSWERS TO EXERCISES

1.

63	48	20	9	72
0	42	81	25	32
56	64	54	28	45
49	30	56	18	8
40	21	36	0	24
7	36	27	35	0

2.
    a. 4,902   b. 5,135   c. 2,204   d. 3,315

3.
    a. 406,224   b. 23,790   c. 28,140   d. 375,737

4.
    a. 12,500   b. 70,300,000   c. 37,000   d. 914,000,000

5.
    a. 314,750   b. 287,348,100

lesson five

DIVISION OF WHOLE NUMBERS

When we divide one number into another number, the answer is called the **quotient**. The division of 12 divided by 3 can be written as:

$$\frac{12}{3} \qquad 12 \div 3 \qquad 3\overline{)12}$$

Quotient

In order to divide, we must know the multiplication facts. Actually, every multiplication problem involves two related division facts. For example, if we know $3 \times 4 = 12$; we then know that $12 \div 4 = 3$ and $12 \div 3 = 4$. Division is called the **inverse** of multiplication.

Multiplicative Inverse

Division can also be thought of as repeated subtraction. For example, $12 \div 3$ can be found by asking how many times 3 can be subtracted from 12.

$12 - 3 = 9$	$9 - 3 = 6$	$6 - 3 = 3$	$3 - 3 = 0$
1 time	2 times	3 times	4 times

Since 3 can be subtracted from 12, 4 times; $12 \div 3 = 4$. We say that there are four multiples of 3 in 12. However, it is easier to do the division $12 \div 3 = 4$ by remembering the multiplication fact $3 \times 4 = 12$.

To find the quotient of larger numbers, we could do repeated subtraction, but it would take too much time. The following problem shows how we can use an **algorithm** (a mechanical process) to do division. We want to divide $3\overline{)1,728}$ (see the next page). The answer of 576 tells us we would have to subtract 3 from 1,728 a total of 576 times.

Division Algorithm

Lesson Five–Division of Whole Numbers

41

Study the example below carefully before going on.

$$\begin{array}{r} 0576 \\ 3{\overline{\smash{\big)}\,1728}} \\ \underline{15} \\ 22 \\ \underline{21} \\ 18 \\ \underline{18} \\ 0 \end{array}$$

In the next example, we would have to subtract 82 from 5,986 a total of 73 times. Using an algorithm 5,986 ÷ 82 is solved as follows:

$$\begin{array}{r} 0073 \\ 82{\overline{\smash{\big)}\,5986}} \\ \underline{574} \\ 246 \\ \underline{246} \\ 0 \end{array}$$

Divisor
Dividend

Let's look at each step. 82 is called the **divisor**; 5,986 is called the **dividend**.

$$\begin{array}{r} 0 \\ 82{\overline{\smash{\big)}\,5986}} \end{array}$$

82 will not go into 5. We put a zero above the 5.

$$\begin{array}{r} 00 \\ 82{\overline{\smash{\big)}\,5986}} \end{array}$$

Trial Divisor

82 will not go into 59. So, we put one more 0 in the answer. Now, 82 will go into 598. Because most of us do not memorize the multiplication table up to the 82's, we use the 8 of the 82 as a **trial divisor** into the 59 of 598. This trial divisor process gives us a close guess to try for an answer. Since 8 × 7 = 56, let's try 7. Put the 7 over the 8 in the dividend.

$$\begin{array}{r} 007 \\ 82{\overline{\smash{\big)}\,5986}} \end{array}$$

If you want, you can now erase the 0 digits before the 7. They did help us put the 7 in the right spot, but we usually do not keep the 0 digits before the first digit. The 7 is now multiplied times 82. The number we get when we multiply 7×82 must be smaller than 598, and it is. Subtract 574 from 598.

$$\begin{array}{r} 73 \\ 82{\overline{\smash{\big)}\,5986}} \\ \underline{574} \\ 24 \end{array}$$ ← partial remainder

Partial Remainder

We used 0 digits to help place the 7 in the answer. But, we could do the problem using our knowledge of place value. We place the 7 in the tens position of 5,986 because $70 \times 82 = 5,740$. 5,740 is smaller than 5,986 and our remainder is smaller than 82. If the 7 had been placed in a different position, we would have too much or too little. For example, if the 7 went over the 9 in the hundreds position of 5,986, we would have $700 \times 82 = 57,400$ which is larger than 5,986.

The 6 must now be brought down to give 246. Now we ask how many times 82 will go into 246. The 8 in 82 is divided this time into 24. The answer, 3, is written after the 7. So, 3 is written above the 6 in the dividend. The 3 is then multiplied by 82 as follows:

$$\begin{array}{r} 73 \\ 82{\overline{\smash{\big)}\,5986}} \\ \underline{574}\downarrow \\ 246 \\ \underline{246} \\ 000 \end{array}$$

Notice this time when we subtracted, the partial remainder was 0, and we are done with the problem.

Lesson Five–Division of Whole Numbers

We Check Our Answer Using Multiplication

We can check our answer by multiplying the quotient, 73, times the divisor, 82.

$$\begin{array}{r} 82 \\ \times\ 73 \\ \hline 246 \\ 574 \\ \hline 5,986 \end{array}$$

The next example shows how to handle zeros in the dividend. We will do the following division.

Handling Zeros In The Dividend

$$69\overline{)35,071}$$

69 will not divide into 3 or 35, so the first digit of our answer will need to go over the 0 in the dividend. This time we will use 7 as a trial divisor, because 69 is so close to 70. We find that 7 goes into 35; giving us a 5, which we write above the 0. We then multiply 5×69 as follows:

$$\begin{array}{r} 005 \\ 69\overline{)35071} \\ 345 \\ \hline 5 \end{array} \leftarrow \text{partial remainder}$$

Notice that the 5 in the quotient represents 500 and $500 \times 69 = 34,500$. Again, we only write down 345. The 7 is now brought down and we try to divide 69 into 57. No luck. So, we put a 0 above the 7 and bring down the 1 as follows:

$$\begin{array}{r} 0050 \\ 69\overline{)35071} \\ 345\downarrow \\ \hline 57 \\ 00\downarrow \\ \hline 571 \end{array}$$

44   Developmental Arithmetic–A Computational Review

Note: We first tried to divide 69 into 57. The zero is not written until we have tried to divide. We then write in the zero in the answer before bringing down the 1.

We now want to divide 69 into 571. The trial divisor of 7 divides into 57 about 8 times. We place the 8 above the 1 and multiply 8 × 69 to get 552. The remainder for the problem is 19.

$$\begin{array}{r} 508 \\ 69{\overline{\smash{)}}\,35071} \\ \underline{345} \\ 571 \\ \underline{552} \\ 19 \leftarrow \text{remainder} \end{array}$$

For the time being, we will not write the remainder as part of our answer. When we get to the review of decimal fractions, we will simply continue the division process by putting in a decimal point and adding some zeros.

Before we finish our review of division, there is one final point.

| **We cannot divide by zero.** |

Note: We Cannot Divide By Zero

We cannot do a problem like $0{\overline{\smash{)}}\,5}$. Whatever answer we would get would have to check. That is, 0 multiplied by our answer would have to equal 5. This is not possible because 0 multiplied by any number is 0, not the 5 we should get in the dividend when we check our answer.

In contrast, it is important to note that we can always divide a non-zero number into zero. We will always get an answer of zero. For example,

$$7{\overline{\smash{)}}\,0} = 0 \text{ because } 0 \times 7 = 0.$$

Note: Zero Divided By Any Number, (≠ 0), Is Zero

Lesson Five—Division of Whole Numbers

REVIEW

See page 41

- **Division**
  Division is the inverse of multiplication. The answer is called the quotient, which we get by dividing the divisor into the dividend.

See page 41

- **Quotient**
  The answer to a division problem is called the quotient.

See page 42

- **The Division Algorithm:**
  1. Decide how many times the divisor will go into the first digits of the dividend. Zeros help keep digits in the correct position according to place value. ($8\overline{)46} \Rightarrow 05$)
  2. Multiply the answer times the divisor. ($5 \times 8 = 40$)
  3. Subtract to get the partial remainder. ($46 - 40 = 6$)
  4. Bring down the next digit in the dividend.
     (Bring down the 4 to make 64.)
  5. Repeat the process as often as necessary.
     ($8\overline{)64} = 8$; $8 \times 8 = 64$; $64 - 64 = 0$; answer 58.)

$$\begin{array}{r} 058 \\ 8\overline{)464} \\ \underline{40}\downarrow \\ 64 \\ \underline{64} \\ 0 \end{array}$$

See page 42

- **Trial divisor**
  We can simplify the process of division by using smaller numbers as divisors; we use the first digit of the divisor rather than all of the digits in the divisor.

See page 45

- **Division by Zero**
  We cannot divide by zero. But zero can be divided by any other number (not equal to zero); the result is zero.

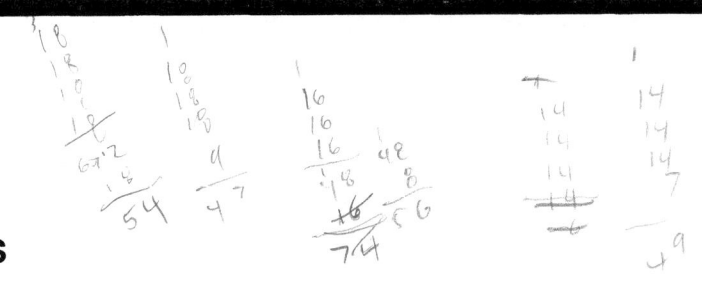

EXERCISES

1. Do the following division as quickly and accurately as possible.

$\dfrac{21}{3} = 7$ $9\overline{)54}$ (6) $56 \div 8 = 7$ $\dfrac{0}{5} = 0$ $7\overline{)49}$ (7)

$8\overline{)72}$ (9) $\dfrac{35}{7} = 5$ $9\overline{)45}$ (5) $42 \div 6 = 7$ $\dfrac{27}{3} = 9$

$48 \div 6 = 8$ $7\overline{)63}$ (9) $\dfrac{56}{7} = 8$ $6\overline{)36}$ (6) $64 \div 8 = 8$

$\dfrac{36}{9} = 4$ $28 \div 7 = 4$ $1\overline{)9}$ (9) $\dfrac{81}{9} = 9$ $8\overline{)56}$ (7)

$3\overline{)24}$ (8) $\dfrac{15}{5} = 3$ $0 \div 7 = 0$ $4\overline{)20}$ (5) $\dfrac{8}{1} = 8$

$25 \div 5 = 5$ $7\overline{)42}$ (6) $\dfrac{100}{10} = 10$ $72 \div 9 = 8$ $8\overline{)56}$ (7)

Work For Accuracy And Quickness

Lesson Five—Division of Whole Numbers   47

2. Find the following quotients:

a. 5)675

b. 8)4,624

c. 47)235

d. 97)8,633

3. Find the following quotients:

a. 2)6,748

b. 3)9,027

c. 5)50,025

d. 8)16,408

4. Divide:
   a. $4{,}212 \div 12$

   b. $73\overline{)13{,}797}$

   c. $\dfrac{20{,}454}{21}$

   d. $69{,}954 \div 89$

5. Divide:
   a. $45{,}034 \div 89$

   b. $40\overline{)2{,}824{,}320}$

   c. $87\overline{)348{,}522}$

   d. $355{,}416 \div 502$

Lesson Five—Division of Whole Numbers

ANSWERS TO EXERCISES

1.

7	6	7	0	7
9	5	5	7	9
8	9	8	6	8
4	4	9	9	7
8	3	0	5	8
5	6	10	8	7

2. a. 135 b. 578 c. 5 d. 89

3. a. 3,374 b. 3,009 c. 10,005 d. 2,051

4. a. 351 b. 189 c. 974 d. 786

5. a. 506 b. 70,608 c. 4,006 d. 708

lesson six

PRIME NUMBERS AND FACTORING

We will use what are called **prime numbers** in our work with fractions. The numbers 2, 3, 5, 7, 11, 13, 17, 19, and 23 are the prime numbers between 1 and 25. It is important to note that 1 is not considered a prime number, and that there are infinitely many primes.

> **Prime numbers can only be written as a product of 1 and the number itself. Note that a prime is always greater than 1.**

Prime Numbers

Prime numbers, therefore, <u>only</u> <u>have</u> <u>two</u> <u>divisors</u>; 1 and the number itself. For example 5 and 17 are prime numbers.

$$5 = 1 \times 5 \qquad 17 = 1 \times 17$$

The numbers that are not prime are called **composite numbers**. Composite numbers have <u>more</u> <u>than</u> <u>two</u> <u>divisors</u> and can be written as a product of two numbers excluding the factor 1. For example 10, 15, and 18 are composite numbers.

Composite Numbers

$$10 = 2 \times 5 \qquad 15 = 3 \times 5 \qquad 18 = 3 \times 6$$
$$18 = 2 \times 9$$

Please note that composite numbers have more than two divisors. The following table shows the divisors of the composite numbers 10, 15, and 18. Each divisor is shown using a division problem.

Lesson Six–Prime Numbers and Factoring

Table of Divisors for the Composite Numbers 10, 15, and 18

$10 \div 2 = 5$	$15 \div 3 = 5$	$18 \div 2 = 9$
$10 \div 5 = 2$	$15 \div 5 = 3$	$18 \div 3 = 6$
$10 \div 10 = 1$	$15 \div 15 = 1$	$18 \div 6 = 3$
$10 \div 1 = 10$	$15 \div 1 = 15$	$18 \div 9 = 2$
		$18 \div 18 = 1$
		$18 \div 1 = 18$

Every Composite Can be Written as a Product of Primes

A very important statement in mathematics tells us that every composite number can be written as a product of prime numbers. When we write a composite number as a product of primes, it is called **prime factoring**. For example the composite numbers 18, 15, and 20 can be written as a product of primes as follows:

$$18 = 2 \times 3 \times 3 = 2^1 \times 3^2$$
$$15 = 3 \times 5 = 3^1 \times 5^1$$
$$20 = 2 \times 2 \times 5 = 2^2 \times 5^1$$

Using Exponents To Tell How Many Times A Factor Is Used

Please note that we used what is called an **exponent** as a shortcut to tell how many times each prime factor was written in the product of primes. The exponent we see here is used in the same way that we used exponents for powers of ten in expanded notation in Lesson 1 and as a shortcut in multiplication by powers of 10 in Lesson 4.

In the prime factoring of 18 we obtained $3^2 \times 2^1$. In the factor 3^2, the exponent 2 tells us that the base 3 was written twice in the product of primes. In the factor 2^1, the exponent 1 tells us that the base 2 was only written once in the product.

We will use exponents to write all of the products we find doing prime factoring. Later on, we will use prime factoring and exponents to find the least common denominator used to add and subtract fractions.

As we work with fractions, we need to factor composite numbers into products of primes. We can often factor a composite number into a product of primes simply by looking at it. For example, writing 15 as the product 3×5 comes easily because we know the multiplication table.

Sometimes we can use **divisibility rules** to help us factor a composite number into a product of primes. The following three divisibility rules should be memorized.

- **Even numbers are divisible by** 2.
- **Numbers that end in a 0 or a 5 are divisible by** 5.
- **Numbers whose digits add up to a multiple of 3 are divisible by** 3

Divisibility Rules

We can use the composite number 150 as an example of using the divisibility rules to write 150 as a product of primes. In the next example on the following page we will use a more structured division process.

Prime Factoring Using the Divisibility Rules

1. Since 150 is an even number, it is divisible by 2. We start by writing $150 = 2 \times 75$.

2. Next, we notice that the digits of 75 add up to 12 which is a multiple of 3. So we can write $75 = 3 \times 25$.

3. If we put the two results together we can write $150 = 2 \times 3 \times 25$.

4. Finally we notice that 25 ends in a 5 so we can write $25 = 5 \times 5$.

5. When we rewrite the previous step with the new result we get the product of primes we want $150 = 2 \times 3 \times 5 \times 5 = 2^1 \times 3^1 \times 5^2$.

However, there are times when we need a more systematic method to do the factoring. The following example helps explain the process. We will factor the composite number 990 into a product of primes.

Lesson Six—Prime Numbers and Factoring

53

Using The Structured Division Process To Get The Prime Factoring of 990

1. We first make a list of the first 11 primes. Using this list, we can prime factor any number under 1,000.

    2, 3, 5, 7, 11, 13, 17, 19, 23, 29, 31

2. We now try dividing, in order, by each prime number in our list of primes. We start by dividing 2 into 990. Note, it helps to use short-division at this point.

    $$\begin{array}{r}495\\2\overline{)990}\end{array}$$

3. Using the divisibility rules, we see that 2 will not divide 495. So we go to the next prime in our list and divide 3 into 495. We note that the sum of the digits in 495 gives a multiple of 3. We can divide by the 3 twice as shown below.

    $$\begin{array}{r}55\\3\overline{)165}\\3\overline{)495}\\2\overline{)990}\end{array}$$

4. The sum of the digits in 55 does not give a multiple of 3; so we go to the next prime in our list of primes. We divide 5 into 55 and get 11. We complete the division by dividing 11 into 11 to get a 1.

    $$\begin{array}{r}1\\11\overline{)11}\\5\overline{)55}\\3\overline{)165}\\3\overline{)495}\\2\overline{)990}\end{array}$$

5. We finish by writing 990 as the product of all the primes we used as divisors. We rewrite the product using exponents as follows:

    $$990 = 2 \times 3 \times 3 \times 5 \times 11 = 2^1 \times 3^2 \times 5^1 \times 11^1$$

We will finish this section with a second example of writing 462 as a product of primes. As you go through the solution, there are several points to notice. We again start by making list of the first 11 primes. We first divide by 2 because 462 is even. Since the sum of the digits in 231 is a multiple of 3, we then divide by 3. After dividing by the 3, we find that the next prime 5 will not divide into 77, so we skip to the next prime which is a 7. We finish by dividing the 11 to get a 1.

$$\begin{array}{r} 1 \\ 11\overline{)11} \\ 7\overline{)77} \\ 3\overline{)231} \\ 2\overline{)462} \end{array}$$

Using The Structured Division Process To Get The Prime Factoring of 462

We now write 462 as a product of primes using exponents.

$$462 = 2^1 \times 3^1 \times 7^1 \times 11^1$$

One last point involves why we divide by the list of primes in order, rather than jumping around. There are times when nothing seems to work, and we do not know if we have gone far enough in our division.

The Importance of Dividing by the Primes in Order

> If we try dividing by the primes in our list, in order, **we can stop at the last prime whose square is still smaller than our number.**

For example if we start with the number 101, we only have to try dividing by the primes 2, 3, 5, and 7. The next prime in the list is 11 and its square is 121; which is larger than 101. Since none of the first four primes will divide exactly into 101, we know that we can stop and say that 101 is a prime number.

When we make our list of primes before we start dividing, we only need to list the primes that have squares smaller than our number. Another way of saying this is that **the only prime divisors we have to try are the ones smaller than the square root of our number.**

Using The Square Root To Write The List Of Prime Divisors

Lesson Six—Prime Numbers and Factoring

REVIEW

See page 51
- **Prime Numbers**

    Prime numbers can only be written as a product of 1 and the number itself. Prime numbers have <u>only</u> two divisors, 1 and the number itself.

See page 51
- **Composite Numbers**

    Composite numbers can be written as the product of two numbers excluding the factor 1. Composite numbers have <u>more than</u> two divisors.

See page 52
- **Prime Factoring**

    Composite numbers can be written as a product of prime numbers. Factoring composite numbers into primes involves dividing by prime numbers until the last quotient is a 1.

See page 52
- **Exponents**

    Exponents are used to show how many times each prime factor is written in the product of primes.

See page 53
- **Divisibility Rules**

    ❖ Even numbers are divisible by 2.

    ❖ Numbers that end in a 0 or a 5 are divisible by 5.

    ❖ Numbers whose digits add up to a multiple of 3 are divisible by 3.

EXERCISES

Use the structured division process shown in this lesson and exponents to write the following composite numbers as a product of primes.

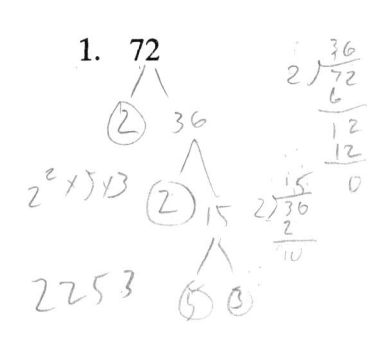

1. 72

2. 420

3. 500

4. 675

5. 336

6. 910

Lesson Six–Prime Numbers and Factoring

57

7. 840   8. 375

9. 1,728   10. 1,311

ANSWERS TO EXERCISES

1. $72 = 2^3 \times 3^2$

2. $420 = 2^2 \times 3^1 \times 5^1 \times 7^1$

3. $500 = 2^2 \times 5^3$

4. $675 = 3^3 \times 5^2$

5. $336 = 2^4 \times 3^1 \times 7^1$

6. $910 = 2^1 \times 5^1 \times 7^1 \times 13^1$

7. $840 = 2^3 \times 3^1 \times 5^1 \times 7^1$

8. $375 = 3^1 \times 5^3$

9. $1,728 = 2^6 \times 3^3$

10. $1,311 = 3^1 \times 19^1 \times 23^1$

Lesson Six–Prime Numbers and Factoring

lesson seven

COMMON FRACTIONS AND MIXED NUMBERS

What is a fraction? Most of us think of a fraction as something like $\frac{1}{2}$ or $\frac{2}{3}$. This kind of fraction is called a **common fraction.** There are two other kinds of fraction that we will study in later lessons; decimal fractions and percents. This lesson will focus on common (or simple) fractions. Decimal fractions will be discussed in Lessons 14 through 17, and percents will be studied in Lesson 18.

Common Fractions

A common fraction is written as one whole number over another whole number. The following examples are common fractions.

$$\frac{3}{1} \quad \frac{1}{2} \quad \frac{7}{4} \quad \frac{17}{20} \quad \frac{2}{3} \quad \frac{8}{5} \quad \frac{0}{6}$$

| **Note: Zero cannot be written at the bottom of a common fraction.** |

The Denominator Cannot Be Zero

We can also look at a common fraction as a division problem. For example, $\frac{2}{3}$ can be written as the division $2 \div 3$ or $3\overline{)2}$. When we get to the review of decimal fractions, we will actually do the division to obtain the decimal fraction for the common fraction $\frac{2}{3}$.

Lesson Seven–Common Fractions and Mixed Numbers　　　　　　　61

Let's look at the common fraction $\frac{1}{2}$ (read as 1 over 2 or one half) to see what we mean by a common fraction. The bottom number, called the **denominator**, tells us that something was broken into 2 equal parts, or pieces of 2. The top number, called the **numerator**, tells us that we have 1 of the pieces of 2. Very young children have difficulty with the idea of a half of anything. One of the great discoveries of childhood is that if we have two halves, we have a whole.

Denominator

Numerator

Most of us can relate to the following anecdote:

A young child comes in from playing a short time before dinner. The child asks for a cookie; but the parent, concerned about dinner, tells the child that he or she can only have half a cookie. The parent then breaks a cookie in two parts and gives one of the pieces of 2 to the child. The child of course wanted a whole cookie, not a half of a cookie. After considerable protest by the child, the parent may relent and hand both halves of the cookie to the child. However, the now bewildered parent finds that the child is not satisfied with the two halves. The child wants a whole cookie. No amount of persuasion will convince the child that the two halves are the same as one whole cookie.

As adults much of our practical experience with fractions still involves food. We measure out ingredients using fractions, and often cut everything from cakes and pies to pizzas into what become fractional parts of the whole.

One of the more interesting social customs involves sandwiches that are cut into bite size pieces, often into 8 triangular pieces. These pieces of 8 are then carried around on a platter, and the hungry guest often finds it necessary to grab a handful or two of the pieces. If our guest only grabs 5 pieces of 8, we say that he or she had five eighths of a whole sandwich. We would write the fraction as $\frac{5}{8}$. A starving guest might grab two handfuls from the plate and end up with 11 pieces of 8, read as eleven eighths. We would write the common fraction as $\frac{11}{8}$.

Common fractions like $\frac{5}{8}$, that have the smaller number on the top are called **proper common fractions**. The following common fractions are examples of proper common fractions.

Proper Common Fractions

$$\frac{1}{32} \qquad \frac{3}{4} \qquad \frac{17}{25} \qquad \frac{2}{3} \qquad \frac{1}{5}$$

The fraction $\frac{11}{8}$ is different from a proper common fraction. The 11 pieces of 8 tells us we have more than one whole sandwich. The numerator of the fraction $\frac{11}{8}$ is larger than the denominator. The fraction $\frac{11}{8}$ is called an **improper common fraction**. Despite what the word improper might suggest, there is nothing wrong with such fractions (the same cannot be said about the behavior of the guest). Here are more examples of improper common fractions.

Improper Common Fractions

$$\frac{3}{2} \qquad \frac{8}{3} \qquad \frac{25}{2} \qquad \frac{4}{1} \qquad \frac{35}{4}$$

We often, but not always, rewrite an improper fraction like $\frac{11}{8}$ as a **mixed number**. A mixed number is a whole number plus a proper

Mixed Numbers

Lesson Seven—Common Fractions and Mixed Numbers 63

fraction. But we do not use a plus sign between the whole number and the fraction part when we write a mixed number. For example we rewrite the common fraction $\frac{11}{8}$ as the mixed number $1\frac{3}{8}$. Please note this is not what we call reducing a fraction. The fraction $\frac{11}{8}$ was already reduced; we just rewrote the improper fraction as the mixed number $1\frac{3}{8}$.

We use division to rewrite improper fractions as mixed numbers.

Writing Improper Fractions As Mixed Numbers

$$\begin{array}{r} 1\frac{3}{8} \\ 8\overline{)11} \\ \underline{8} \\ 3 \end{array}$$

The remainder of 3 tells us that we have 3 pieces of 8. We write the remainder over the divisor 8 to get the proper fraction $\frac{3}{8}$. We then write the proper fraction after the 1 in the quotient to get the mixed number $1\frac{3}{8}$.

The mixed number tells us more quickly how many wholes we have. For this reason, mixed numbers are often preferred over the improper fraction. Even though at some point in the past you may have been told to never leave an improper fraction as an answer, you need to be comfortable with either form. To help you become more comfortable with improper fractions, the answers to the exercises in this book are given as both improper fractions and mixed numbers. The following examples are improper fractions that have been written as mixed numbers:

$$\frac{5}{3} = 1\frac{2}{3} \qquad \frac{3}{2} = 1\frac{1}{2} \qquad \frac{17}{5} = 3\frac{2}{5}$$

Being able to change a mixed number back into an improper fraction is important; and it is easy to do. Let's look again at the mixed number $1\frac{3}{8}$. The 1 can be written as 8 pieces of 8. We would then have 8 + 3 or 11 pieces of 8.

Writing Mixed Numbers As Improper Fractions

Here is a rule that is used to change mixed numbers into improper fractions. Multiply the whole number (a 1 in this example) by the denominator of the fraction (an 8 this time) and add the numerator of the fraction (a 3). The result is the numerator of the improper fraction. The improper fraction will have the same denominator as the fraction part of the mixed number (an 8 in our example).

Mixed Numbers to Fractions

$$1\frac{3}{8} = \frac{8 \times 1 + 3}{8} = \frac{11}{8}$$

Let's look at a second example to make sure the process is illustrated well enough.

$$2\frac{1}{4} = \frac{4 \times 2 + 1}{4} = \frac{9}{4}$$

Lowest Terms

It is usual to write common fractions in **lowest terms**. For example the fraction $\frac{2}{3}$ is in lowest terms because the only number that will divide into both the numerator and the denominator is 1. The

Lesson Seven—Common Fractions and Mixed Numbers 65

Reducing Fractions To Lowest Terms

fraction $\frac{6}{8}$ is not in lowest terms because 2 will divide evenly into both the numerator and the denominator. If the numerator and denominator are both divisible by the same number (excluding 1), then the fraction is not in lowest terms. We can reduce $\frac{6}{8}$ to $\frac{3}{4}$ by dividing 2 into both the numerator and denominator. We say that we used cancelling to write the fraction in lowest terms. We call the process **reducing a fraction**.

In this example it was easy to see the largest number that would divide into both the numerator and denominator. But there are times when it is very hard to find the largest number that will go into both parts of the fraction. In such cases we often try dividing by smaller numbers, continuing until we have the fraction reduced. Unfortunately, it is very easy to make errors when we do cancelling.

Using Prime Factoring To Reduce Fractions

There is another method to reduce fractions. We can use prime factoring to reduce common fractions to lowest terms. We will look at the process using the same fraction, $\frac{6}{8}$.

Step 1.

$$\text{numerator} \qquad \text{denominator}$$

$$\begin{array}{c} 3\overline{)3}^{\,1} \\ 2\overline{)6} \end{array} \quad 6 = 2 \times 3 \qquad \begin{array}{c} 2\overline{)2}^{\,1} \\ 2\overline{)4} \\ 2\overline{)8} \end{array} \quad 8 = 2 \times 2 \times 2$$

Step 2. $\quad \dfrac{6}{8} = \dfrac{\overset{1}{2} \times 3}{\underset{1}{2} \times 2 \times 2}$

Developmental Arithmetic–A Computational Review

Step 3. $\dfrac{6}{8} = \dfrac{1 \times 3}{1 \times 2 \times 2} = \dfrac{3}{4}$

The three steps to reduce a fraction to lowest terms using prime factoring:

Step 1. Factor both the numerator and denominator into products of primes. Do not use exponents this time.

Step 2. Rewrite the fraction in factored form and cross out **like factors**. Remember to write ones above the crossed out numbers.

Step 3. Multiply the remaining factors together in the numerator.

The Three Steps To Reduce A Fraction Using prime Factoring

Crossing Out Like Factors

We will look at a second example of using prime factoring to reduce the common fraction $\dfrac{30}{75}$.

Step 1.

$$\begin{array}{r}1\\5\overline{)5}\\3\overline{)15}\\2\overline{)30}\end{array} \quad 30 = 2 \times 3 \times 5 \qquad \begin{array}{r}1\\5\overline{)5}\\5\overline{)25}\\3\overline{)75}\end{array} \quad 75 = 3 \times 5 \times 5$$

Step 2.

$$\dfrac{30}{75} = \dfrac{2 \times \overset{1}{\cancel{3}} \times \overset{1}{\cancel{5}}}{\underset{1}{\cancel{3}} \times \underset{1}{\cancel{5}} \times 5}$$

Step 3.

$$\dfrac{30}{75} = \dfrac{2 \times 1 \times 1}{1 \times 1 \times 5} = \dfrac{2}{5}$$

Lesson Seven—Common Fractions and Mixed Numbers

REVIEW

See page 63

- **Proper Common Fractions**
  The numerator of the common fraction is smaller than the denominator.

See page 63

- **Improper Common Fractions**
  The numerator of the common fraction is larger than the denominator. Improper fractions can be changed to mixed numbers by dividing the numerator by the denominator and writing the remainder as a fraction.

See page 63

- **Mixed Numbers**
  A mixed number is the sum of a whole number and a fraction written without the plus sign. Mixed numbers can be changed to improper fractions by multiplying the whole number by the denominator, adding the numerator, and putting the result over the denominator.

See page 67

- **Reducing Common Fractions**
  Common fractions can be reduced to lowest terms by writing the numerator and denominator as products of primes, crossing out like factors, and then multiplying the remaining factors together.

See page 65

- **Lowest Terms**
  A fraction is in lowest terms when 1 is the only divisor that will go into both the numerator and denominator.

EXERCISES

1. Use the letters P (Proper), I (Improper), and M (Mixed) to classify each of the fractions below as proper fractions, improper fractions, or mixed numbers.

    a. $\dfrac{5}{2}$ _I_   b. $1\dfrac{1}{4}$ _M_   c. $\dfrac{2}{3}$ _P_

    d. $\dfrac{1}{4}$ _P_   e. $\dfrac{3}{2}$ _I_   f. $2\dfrac{1}{3}$ _M_

    g. $4\dfrac{1}{2}$ _M_   h. $\dfrac{2}{5}$ _P_   i. $\dfrac{5}{4}$ _I_

2. Rewrite each of the following improper fractions as a mixed number.

    a. $\dfrac{7}{3}$ $2\dfrac{1}{3}$ b. $\dfrac{3}{2}$ $1\dfrac{1}{2}$ c. $\dfrac{5}{4}$ $1\dfrac{1}{4}$

    d. $\dfrac{17}{5}$ $3\dfrac{2}{5}$ e. $\dfrac{7}{2}$ $3\dfrac{1}{2}$ f. $\dfrac{23}{4}$ $5\dfrac{3}{4}$

    g. $\dfrac{19}{8}$ 2 h. $\dfrac{13}{3}$ $4\dfrac{1}{3}$ i. $\dfrac{23}{7}$ $3\dfrac{2}{7}$

Lesson Seven—Common Fractions and Mixed Numbers

3. Change each of the mixed numbers to an improper fraction.

a. $1\frac{1}{2}$ $\frac{3}{2}$

b. $2\frac{1}{3}$ $\frac{7}{3}$

c. $4\frac{1}{2}$ $\frac{9}{2}$

d. $3\frac{1}{2}$ $\frac{7}{2}$

e. $2\frac{3}{8}$ $\frac{19}{8}$

f. $3\frac{1}{4}$ $\frac{13}{4}$

g. $1\frac{1}{5}$ $\frac{6}{5}$

h. $5\frac{1}{3}$ $\frac{16}{3}$

i. $9\frac{2}{7}$ $\frac{67}{7}$

4. Use prime factoring to reduce each of the following fractions.

a. $\dfrac{12}{15}$ $\dfrac{3}{5}$

b. $\dfrac{6}{18}$ $\dfrac{3}{9}$

c. $\dfrac{6}{4}$ $\dfrac{3}{2}$

d. $\dfrac{25}{10}$ $\dfrac{5}{2}$

e. $\dfrac{18}{24}$ $\dfrac{3}{4}$

f. $\dfrac{10}{12}$ $\dfrac{5}{6}$

g. $\dfrac{30}{45}$ $\dfrac{2}{3}$

h. $\dfrac{36}{84}$ $\dfrac{3}{7}$

Lesson Seven—Common Fractions and Mixed Numbers

ANSWERS TO EXERCISES

1.
 - a. I
 - b. M
 - c. P
 - d. P
 - e. I
 - f. M
 - g. M
 - h. P
 - i. I

2.
 - a. $2\frac{1}{3}$
 - b. $1\frac{1}{2}$
 - c. $1\frac{1}{4}$
 - d. $3\frac{2}{5}$
 - e. $3\frac{1}{2}$
 - f. $5\frac{3}{4}$
 - g. $2\frac{3}{8}$
 - h. $4\frac{1}{3}$
 - i. $3\frac{2}{7}$

3.
 - a. $\frac{3}{2}$
 - b. $\frac{7}{3}$
 - c. $\frac{9}{2}$
 - d. $\frac{7}{2}$
 - e. $\frac{19}{8}$
 - f. $\frac{13}{4}$
 - g. $\frac{6}{5}$
 - h. $\frac{16}{3}$
 - i. $\frac{65}{7}$

4.
 - a. $\frac{4}{5}$
 - b. $\frac{1}{3}$
 - c. $\frac{3}{2}$
 - d. $\frac{5}{2}$
 - e. $\frac{3}{4}$
 - f. $\frac{5}{6}$
 - g. $\frac{2}{3}$
 - h. $\frac{3}{7}$

lesson eight
THE BASIC PRINCIPLE OF FRACTIONS

In order to add or subtract common fractions, it is necessary for the fractions to have the same denominator. When fractions have the same denominator, we say they have a **common denominator**. For example, we can add $\frac{1}{3}$ and $\frac{2}{3}$ because the fractions have a common denominator of 3.

Common Denominator

The fraction $\frac{1}{3}$ tells us that we have 1 piece of something that was broken into 3 equal parts. The fraction $\frac{2}{3}$ tells us that we have 2 pieces of something that was broken into 3 equal parts. We add the "1 piece of 3" to the "2 pieces of 3" to get "3 pieces of 3" or 1 as shown below.

$$\frac{1}{3} + \frac{2}{3} = \frac{1+2}{3} = \frac{3}{3} = 1$$

Adding Fractions

> **Remember:** We must have a common denominator before we can add or subtract common fractions.

Lesson Eight–The Basic Principle of Fractions

You may be more used to writing the addition we just completed in the vertical arrangement as follows:

$$\begin{array}{r}\dfrac{1}{3}\\+\dfrac{2}{3}\\\hline\dfrac{3}{3}\end{array}$$

Adding Fractions

We will look at one more example; we will add $\dfrac{2}{7}$ and $\dfrac{3}{7}$.

$$\frac{2}{7}+\frac{3}{7}=\frac{2+3}{7}=\frac{5}{7}$$

or

$$\begin{array}{r}\dfrac{2}{7}\\+\dfrac{3}{7}\\\hline\dfrac{5}{7}\end{array}$$

The rule for adding fractions that have common denominators is:

Rule For Adding Fractions With A Common Denominator

Add the numerators and put the sum over the common denominator.

If two fractions do not have a common denominator, we cannot add the fractions until we change the denominators of the fractions to a common denominator. For example, we cannot add $\dfrac{1}{2}$ and $\dfrac{1}{3}$. These two fractions do not have a common denominator.

74   Developmental Arithmetic—A Computational Review

We will have to change the fractions to new fractions with the same denominator, called the least common denominator. Then we will be able to add the two fractions. Adding fractions that do not have common denominators is studied in the next lesson.

It is just as easy to subtract fractions with common denominators as it is to add them. As an example, we will do the subtraction

$$\frac{3}{5} - \frac{1}{5}$$

or as it is written in the more familiar form

$$\begin{array}{r}\frac{3}{5}\\-\frac{1}{5}\\\hline\end{array}$$

Subtracting Fractions

We subtract the numerators and put the answer over the common denominator as shown below.

$$\frac{3}{5} - \frac{1}{5} = \frac{3-1}{5} = \frac{2}{5}$$

Remember: If two fractions do not have a common denominator, **we cannot add or subtract the fractions until we change them to new fractions with a common denominator.** Changing fractions to new fractions with a different denominator is important to the process of

Caution!

Lesson Eight–The Basic Principle of Fractions

adding and subtracting fractions. So we will spend the rest of this lesson on the topic of changing fractions.

We are now going to see how we change a fraction into an equal fraction that has a different denominator. For example, the fraction $\frac{1}{2}$ can be written as an equal fraction with a denominator of 6. We set up the problem as follows:

$$\frac{1}{2} = \frac{?}{6}$$

Changing To A New Denominator

Our problem is to find the missing numerator.

Step 1. Divide the new denominator 6 by the original denominator 2.

$$6 \div 2 = 3$$

Step 2. The answer of 3 tells us that the original denominator needs to be multiplied by 3 in order to get the new denominator of 6.

$$\frac{}{2 \times 3} = \frac{}{6}$$

Step 3. Therefore, we multiply the original numerator 1 by the same 3 we found in step 2. This will give us the new fraction with a denominator of 6 equal to the original fraction with a denominator of 3.

$$\frac{1 \times 3}{2 \times 3} = \frac{3}{6}$$

Remember: When we change a fraction to a new fraction with a different denominator, we must use the basic principle of fractions. Our new fraction will then be equal to the original fraction.

The Basic Principle of Fractions:

| **If we multiply (or divide) both the numerator and the denominator of a common fraction by the same number (not zero), the new fraction will equal the original fraction.** |

The Basic Principle Of Fractions

Let's look at a second example. This time we will change the common fraction $\frac{2}{3}$ to an equal fraction with the denominator 12.

$$\frac{2}{3} = \frac{?}{12}$$

We divide the 12 by the 3.

$$12 \div 3 = 4$$

This tells us that we must multiply both the numerator and denominator by 4.

$$\frac{2 \times 4}{3 \times 4} = \frac{8}{12}$$

The basic principle of fractions tells us the new fraction $\frac{8}{12}$ is equal to the original fraction $\frac{2}{3}$. We can always check by reducing the new fraction. We should get the original fraction for our answer.

$$\frac{8}{12} = \frac{\overset{1}{\cancel{2}} \times \overset{1}{\cancel{2}} \times 2}{\underset{1}{\cancel{2}} \times \underset{1}{\cancel{2}} \times 3} = \frac{2}{3}$$

Lesson Eight—The Basic Principle of Fractions

REVIEW

See page 73

- **Common Denominator**
  When two or more fractions have the same denominator, we say they have a common denominator.

See page 74

- **Adding Fractions**
  To add fractions with common denominators, we add the numerators and put the sum over the common denominator.

See page 75

- **Subtracting Fractions**
  To subtract fractions with common denominators, we subtract the numerators and put the difference over the common denominator.

See page 76

- **Changing Fractions**
  Changing a common fraction to an equal fraction with a different denominator requires multiplying both parts of the fraction by the same number.

See page 77

- **Basic Principle of Fractions**
  If we multiply or divide the denominator and numerator of a common fraction by the same number, we will get a fraction that is equal to the original.

EXERCISES

1. Use the Basic Principle of Fractions to change each fraction to a new fraction with the denominator specified.

   a. $\dfrac{1}{3} = \dfrac{}{12}$ b. $\dfrac{2}{5} = \dfrac{}{20}$ c. $\dfrac{3}{4} = \dfrac{}{12}$

   d. $\dfrac{5}{8} = \dfrac{}{24}$ e. $\dfrac{3}{7} = \dfrac{}{21}$ f. $\dfrac{5}{12} = \dfrac{}{36}$

   g. $\dfrac{7}{2} = \dfrac{}{6}$ h. $\dfrac{16}{7} = \dfrac{}{21}$ i. $\dfrac{9}{5} = \dfrac{}{20}$

2. Add the following common fractions.

   a. $\dfrac{1}{5} + \dfrac{2}{5}$ b. $\dfrac{3}{8} + \dfrac{2}{8}$ c. $\dfrac{5}{9} + \dfrac{1}{9}$

Lesson Eight–The Basic Principle of Fractions

d. $\dfrac{2}{6}+\dfrac{4}{6}$ e. $\dfrac{3}{10}+\dfrac{5}{10}$ f. $\dfrac{2}{7}+\dfrac{3}{7}$

g. $\dfrac{3}{11}+\dfrac{6}{11}$ h. $\begin{array}{r}\dfrac{5}{12}\\[4pt]+\ \dfrac{7}{12}\\ \hline\end{array}$ i. $\dfrac{7}{4}+\dfrac{3}{4}$

3. Subtract the following common fractions.

a. $\begin{array}{r}\dfrac{3}{4}\\[4pt]-\ \dfrac{1}{4}\\ \hline\end{array}$ b. $\begin{array}{r}\dfrac{5}{8}\\[4pt]-\ \dfrac{3}{8}\\ \hline\end{array}$ c. $\begin{array}{r}\dfrac{3}{5}\\[4pt]-\ \dfrac{2}{5}\\ \hline\end{array}$

d. $\dfrac{7}{10} - \dfrac{3}{10}$ e. $\dfrac{5}{7} - \dfrac{2}{7}$ f. $\dfrac{5}{6} - \dfrac{1}{6}$

g. $\dfrac{11}{18} - \dfrac{7}{18}$ h. $\begin{array}{r} \dfrac{13}{20} \\ - \dfrac{3}{20} \\ \hline \end{array}$ i. $\dfrac{5}{12} - \dfrac{3}{12}$

Lesson Eight—The Basic Principle of Fractions

ANSWERS TO EXERCISES

1.
   a. $\dfrac{1\times 4}{3\times 4}=\dfrac{4}{12}$ b. $\dfrac{2\times 4}{5\times 4}=\dfrac{8}{20}$ c. $\dfrac{3\times 3}{4\times 3}=\dfrac{9}{12}$

   d. $\dfrac{5\times 3}{8\times 3}=\dfrac{15}{24}$ e. $\dfrac{3\times 3}{7\times 3}=\dfrac{9}{21}$ f. $\dfrac{5\times 3}{12\times 3}=\dfrac{15}{36}$

   g. $\dfrac{7\times 3}{2\times 3}=\dfrac{21}{6}$ h. $\dfrac{16\times 3}{7\times 3}=\dfrac{48}{21}$ i. $\dfrac{9\times 4}{5\times 4}=\dfrac{36}{20}$

2.
   a. $\dfrac{3}{5}$ b. $\dfrac{5}{8}$ c. $\dfrac{6}{9}=\dfrac{2}{3}$

   d. $\dfrac{6}{6}=1$ e. $\dfrac{8}{10}=\dfrac{4}{5}$ f. $\dfrac{5}{7}$

   g. $\dfrac{9}{11}$ h. $\dfrac{12}{12}=1$ i. $\dfrac{10}{4}=\dfrac{5}{2}$

3.
   a. $\dfrac{2}{4}=\dfrac{1}{2}$ b. $\dfrac{2}{8}=\dfrac{1}{4}$ c. $\dfrac{1}{5}$

   d. $\dfrac{4}{10}=\dfrac{2}{5}$ e. $\dfrac{3}{7}$ f. $\dfrac{4}{6}=\dfrac{2}{3}$

   g. $\dfrac{4}{18}=\dfrac{2}{9}$ h. $\dfrac{10}{20}=\dfrac{1}{2}$ i. $\dfrac{2}{12}=\dfrac{1}{6}$

lesson nine
ADDITION OF FRACTIONS

Let's review again what we mean by the fraction $\frac{1}{2}$. Suppose we take a small pizza and cut it into 2 equal pieces. If we keep one piece, we say we have 1 of the 2 equal parts or $\frac{1}{2}$ of the pizza.

Reviewing What A Fraction Means

Now suppose someone else took a pizza just like ours and cut it into 4 equal pieces. If they gave us three pieces of their pizza, we would say we have $\frac{3}{4}$ of their pizza.

Assume that we eat our 1 piece of pizza and their 3 pieces of pizza. If we want to know how much pizza we have eaten, we need to add the fractions $\frac{1}{2}$ and $\frac{3}{4}$. We first need to compare the way the pizzas were cut up. Our first pizza was cut into 2 parts. The second pizza was

Lesson Nine–Addition of Fractions   83

cut into 4 parts. If we want to add $\frac{1}{2}$ and $\frac{3}{4}$, the pizzas must be divided into the same number of parts.

Let's see what we can do. If the first pizza had been cut into 4 equal parts, we would have 2 pieces of 4.

Now we can answer the question because each pizza is cut into the same number of pieces. We have $\frac{2}{4}$ of a pizza and $\frac{3}{4}$ of a pizza.

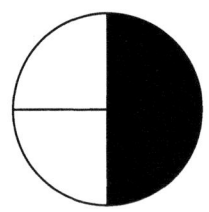

We have a total of 5 pieces. Each piece is $\frac{1}{4}$ of the pizza. So our answer is $\frac{5}{4}$. This is exactly what we have to do if we want to add the fractions $\frac{1}{2}$ and $\frac{3}{4}$.

We could have done the above problem using pieces of 8. Let's try cutting the pizza into smaller pieces for easier handling. We cut it into "pieces of 8". Now $\frac{1}{2}$ of the pizza would give us 4 "pieces of 8" and $\frac{3}{4}$ of the pizza would give us 6 "pieces of 8".

When we add the pieces, we have $\frac{4}{8} + \frac{6}{8}$ or $\frac{10}{8}$ of a pizza. Notice that we now need to reduce $\frac{10}{8}$ to lowest terms.

$$\frac{10}{8} = \frac{\overset{1}{\cancel{2}} \times 5}{\underset{1}{\cancel{2}} \times 2 \times 2} = \frac{5}{4}$$

After we reduce $\frac{10}{8}$ to $\frac{5}{4}$ we get the same answer as before.

When we change fractions to new fractions with "big" denominators, the larger denominators make the problem harder. Usually we choose to use the smallest denominator we can. The process of adding the fractions $\frac{1}{2}$ and $\frac{3}{4}$ required that we first find a common denominator. In this first example it was easy to see that 4 was the smallest common denominator. This was a special case where the larger of the two denominators was the smallest common denominator. **It always pays to check the denominators to see if this shortcut works.**

Sometimes it is not so easy to find the smallest common denominator. We will use the next example to show an easy method that can be used to find the smallest common denominator. This time we will look at what we would do if we were given $\frac{1}{3}$ of one pizza and $\frac{4}{5}$ of

Always Check To See If The Larger Denominator Is The Common Denominator

Lesson Nine—Addition of Fractions

another pizza. Again we want to find out how much pizza we would have. The problem is just like adding the fractions $\frac{1}{3}$ and $\frac{4}{5}$.

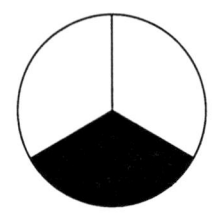

Again, the pieces are not the same size in each pizza. This time it is not as easy to think of a way to cut up the pizzas so that the pieces in both pizzas are all the same size. We need an easy way to find a smallest common denominator. What we are really after is the smallest number that both 3 and 5 will divide into evenly. By inspection we can see that 3 and 5 each divide into 15 evenly. So we could cut the pizzas into "pieces of 15".

$$15 \div 3 = 5 \qquad\qquad 15 \div 5 = 3$$

Rather than relying on inspection to find the smallest common denominator, we can use what is called a list of multiples. We simply write a list of the multiples of each denominator. This counting by the denominators is like we did as kids when we counted by twos and fives.

As an example, we will make a list for each of the denominators of the fractions $\frac{1}{3}$ and $\frac{4}{5}$. We will need a list for the **Multiples of 3** and for the **Multiples of 5**. We can make the lists either by counting by 3s and counting by 5s or by thinking of the lists as a partial multiplication table.

×	Multiples of 3	Multiples of 5
1	3	5
2	6	10
3	9	15
4	12	20
5	15	25

Using A Table Of Multiples

As we continue the list we stop when we find a multiple that is in both lists. We call this multiple the common multiple. Since it is the smallest common multiple, it is referred to as the **least common multiple or LCM**.

The Least Common Multiple

When we look at the two lists of multiples, we find that 15 is the least common multiple or LCM. We will use the LCM as the new denominator for each of the fractions. We need to change $\frac{1}{3}$ and $\frac{4}{5}$ to new fractions with a denominator of 15.

$$\frac{1}{3} = \frac{1 \times 5}{3 \times 5} = \frac{5}{15} \qquad \frac{4}{5} = \frac{4 \times 3}{5 \times 3} = \frac{12}{15}$$

We now have 5 "pieces of 15" and 12 "pieces of 15". We can now add our fractions to get 17 "pieces of 15" or $\frac{17}{15}$ which can be written as the mixed number $1\frac{2}{15}$.

Lesson Nine—Addition of Fractions   87

The least common multiple of the denominators is the method we will use in this lesson to add fractions with different denominators.

The three steps to follow to find the LCM when we add fractions:

Step 1. Write the list of multiples for each denominator until we find the least common multiple, LCM.

Three Steps To Find The LCM

Step 2. Write each fraction with the new denominator using the basic principle of fractions.

Step 3. Add the numerators and put the sum over the LCM.

We will use the three steps to add the following fractions.

$$+\begin{array}{c}\frac{3}{8}\\ \frac{5}{6}\end{array}$$

Step 1.

×	Multiples of 8	Multiples of 6
1	8	6
2	16	12
3	24	18
4	32	24

Step 2. The LCM is 24. We can use the basic principle of fractions to change each fraction to a new fraction with the denominator 24.

$$\frac{3}{8} = \frac{3 \times 3}{8 \times 3} = \frac{9}{24} \qquad \frac{5}{6} = \frac{5 \times 4}{6 \times 4} = \frac{20}{24}$$

Step 3. We add the fractions:

$$\frac{9}{24}+\frac{20}{24}=\frac{29}{24} \text{ or } 1\frac{5}{24}$$

Adding mixed numbers is not much harder to do than adding common fractions. When we add mixed numbers, we just add the fraction parts and add the whole parts. As an example we will find the sum of the fractions $1\frac{1}{3}$ and $2\frac{1}{4}$.

Adding Mixed Numbers

$$1\frac{1}{3}+2\frac{1}{4} \quad \text{or} \quad \begin{array}{r} 1\frac{1}{3} \\ +\ 2\frac{1}{4} \\ \hline \end{array}$$

Step 1. We can see by inspection that the LCM of 3 and 4 is 12.

Step 2. Change the fractions using the basic principle of fractions.

$$1\frac{1}{3}=1\frac{4}{12} \qquad\qquad 2\frac{1}{4}=2\frac{3}{12}$$

Step 3. Add the fractions.

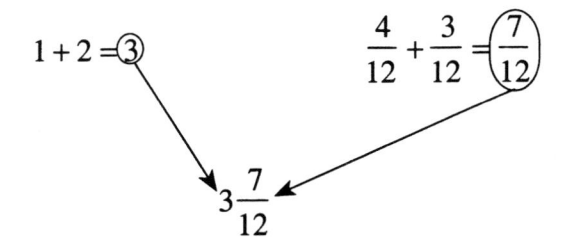

$$1+2=3 \qquad\qquad \frac{4}{12}+\frac{3}{12}=\frac{7}{12}$$

$$3\frac{7}{12}$$

Lesson Nine—Addition of Fractions

What To Do With An Improper Fraction When Adding Mixed Numbers

Sometimes when we add two mixed numbers, the sum of the fraction parts gives us an improper fraction that must be rewritten as a mixed number. The following example shows what we do in such problems. We will find the sum

$$2\frac{3}{5} + 3\frac{1}{2}$$

Step 1. Again by inspection we find the LCM of 10.

Step 2. We change each fraction to a new fraction with the denominator 10.

$$\frac{3}{5} = \frac{3 \times 2}{5 \times 2} = \frac{6}{10} \qquad \frac{1}{2} = \frac{1 \times 5}{2 \times 5} = \frac{5}{10}$$

Step 3. We now add the whole parts and the fraction parts.

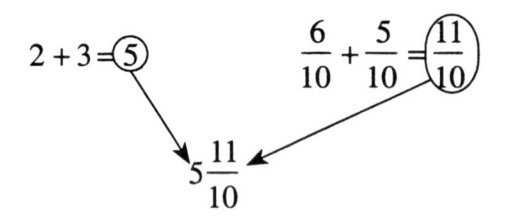

$$2 + 3 = 5 \qquad \frac{6}{10} + \frac{5}{10} = \frac{11}{10}$$

$$5\frac{11}{10}$$

We cannot write a mixed number with the fraction part as an improper fraction. We must change the improper fraction to a mixed number and rewrite the result as follows.

$$5\frac{11}{10} = 5 + 1\frac{1}{10} = 6\frac{1}{10}$$

One way of avoiding the above problem is to first write each mixed number as an improper fraction. We then add the two improper fractions by finding the LCM of each fraction, rewriting the fractions, and adding the numerators over the LCM. The following example shows what is done.

$$2\frac{3}{5} = \frac{13}{5} = \frac{26}{10}$$
$$+ \quad 3\frac{1}{2} = \frac{7}{2} = \frac{35}{10}$$
$$\frac{61}{10} = 6\frac{1}{10}$$

Adding Mixed Numbers By First Changing The Mixed Numbers To Improper Fractions

The idea of first changing mixed numbers to improper fractions will also be used in subtraction of mixed numbers to avoid problems with borrowing. Of course, it will be necessary to change mixed numbers to improper fractions when we both multiply and divide mixed numbers. The point is that the above example illustrates an important idea that goes beyond just being an alternate method for adding mixed numbers.

If we return to the answer we obtained using mixed numbers, we find that we were actually adding a whole number to a mixed number. We could have written our answer as

$$1\frac{1}{10}$$
$$+ \quad 5$$
$$6\frac{1}{10}$$

Adding A Whole Number And A Mixed Number

We can use this result as a guide for adding whole numbers to mixed numbers.

Lesson Nine—Addition of Fractions 91

Adding A Common Fraction To A Mixed Number

Sometimes we want to add a common fraction to a mixed number. Our last example shows how we do the addition.

$$\begin{array}{r} 1\frac{2}{3} \\ + \frac{3}{4} \\ \hline \end{array}$$

Step 1. The LCM is 12.

Step 2.

$$\frac{2}{3} = \frac{2 \times 4}{3 \times 4} = \frac{8}{12} \qquad \frac{3}{4} = \frac{3 \times 3}{4 \times 3} = \frac{9}{12}$$

Step 3.

$$\begin{array}{r} 1\frac{8}{12} \\ + \frac{9}{12} \\ \hline 1\frac{17}{12} \end{array} \qquad 1\frac{17}{12} = 1 + 1\frac{5}{12} = 2\frac{5}{12}$$

Again, we could have avoided the problem in the final step with the improper fraction by first writing $1\frac{2}{3}$ as $\frac{5}{3}$ as shown in the alternate solution below.

$$\begin{array}{r} \frac{5}{3} = \frac{20}{12} \\ + \frac{3}{4} = \frac{9}{12} \\ \hline \frac{29}{12} = 2\frac{5}{12} \end{array}$$

REVIEW

- **Least Common Multiple (LCM)**
  We find the LCM by writing a list of multiples of each denominator and finding the smallest multiple that is common to both lists.

  See page 87

- **Adding Fractions With Different Denominators**
  In order to add fractions that do not have the same denominator, it is necessary to first change each fraction to a new fraction with the LCM as the common denominator. We use the Basic Principle of Fractions to change each fraction. We then add the numerators and place the sum over the LCM.

  See page 88

- **Mixed Numbers Versus Improper Fractions**
  If the answer for the sum of two fractions is an improper fraction we can write our answer as a mixed number, but <u>we do not have to rewrite the answer as a mixed number</u>. We can use improper fractions for the mixed numbers to avoid borrowing when we are subtracting mixed numbers.

  See page 91

Lesson Nine—Addition of Fractions

EXERCISES

Add the following fractions using the appropriate method.

1. $\dfrac{1}{2}$ → $\dfrac{3}{6}$
   $+\dfrac{1}{3}$ → $\dfrac{2}{6}$
   ———
   $\dfrac{5}{6}$

2. $\dfrac{2}{5}$ → $\dfrac{6}{15}$
   $+\dfrac{1}{4}$ → $\dfrac{4}{15}$
   ———
   $\dfrac{10}{15}$ $\dfrac{2}{3}$

3. $\dfrac{2}{3}$ → $\dfrac{8}{12}$
   $+\dfrac{3}{4}$ → $\dfrac{9}{12}$ $\dfrac{18}{12}$
   ———
   $1\dfrac{6}{12}$ or $1\dfrac{1}{2}$

4. $\dfrac{5}{6}$ → $\dfrac{10}{12}$
   $+\dfrac{3}{4}$ → $\dfrac{9}{12}$ $\dfrac{19}{12}$
   ———
   $1\dfrac{7}{12}$

5. $\dfrac{5}{12}$ → $\dfrac{25}{60}$
   $+\dfrac{7}{10}$ → $\dfrac{42}{60}$ $\dfrac{67}{60}$
   ———
   $1\dfrac{7}{60}$

6. $\dfrac{15}{18}$ → $\dfrac{30}{36}$
   $+\dfrac{7}{12}$ → $\dfrac{21}{36}$ $\dfrac{54}{36}$
   ———
   $1\dfrac{18}{36}$ or $1\dfrac{1}{2}$

7. $\dfrac{3}{14} \quad \dfrac{9}{42}$
   $+ \dfrac{8}{21} \quad \dfrac{16}{42}$
   $\dfrac{25}{42}$

8. $\dfrac{5}{27} \quad \dfrac{20}{108}$
   $+ \dfrac{3}{36} \quad \dfrac{9}{108}$
   $\dfrac{29}{108}$

9. $1\dfrac{2}{3} \quad \dfrac{8}{12}$
   $+ 3\dfrac{1}{4} \quad \dfrac{3}{12}$
   $\dfrac{11}{12}$

10. $2\dfrac{1}{3} \quad \dfrac{5}{15}$
    $+ 4\dfrac{3}{5} \quad \dfrac{9}{15}$
    $\dfrac{14}{15}$

11. $3\dfrac{5}{8} \quad \dfrac{10}{16}$
    $+ 2\dfrac{3}{4} \quad \dfrac{12}{16}$
    $6\dfrac{6}{16}$ or $\dfrac{3}{8}$

12. $2\dfrac{15}{16} \quad \dfrac{45}{48}$
    $+ 5\dfrac{7}{12} \quad \dfrac{28}{48}$
    $8\dfrac{25}{48}$

13. $7\dfrac{2}{3} \quad \dfrac{10}{15}$
    $+ \dfrac{3}{5} \quad \dfrac{9}{15}$
    $8\dfrac{4}{15}$ or $8\dfrac{1}{4}$

14. 2
    $+ 3\dfrac{7}{10}$
    $5\dfrac{7}{10}$

Lesson Nine—Addition of Fractions

95

ANSWERS TO EXERCISES

1.

Multiples of 2	Multiples of 3	
2	3	$\frac{1}{2} = \frac{1 \times 3}{2 \times 3} = \frac{3}{6}$
4	6	$+ \frac{1}{3} = \frac{1 \times 2}{3 \times 2} = \frac{2}{6}$
6		$\frac{5}{6}$

2.

Multiples of 5	Multiples of 4	
	4	$\frac{2}{5} = \frac{2 \times 4}{5 \times 4} = \frac{8}{20}$
5	8	$+ \frac{1}{4} = \frac{1 \times 5}{4 \times 5} = \frac{5}{20}$
10	12	$\frac{13}{20}$
15	16	
20	20	

3. $\frac{17}{12}$ or $1\frac{5}{12}$ 4. $\frac{19}{12}$ or $1\frac{7}{12}$ 5. $\frac{67}{60}$ or $1\frac{7}{60}$

6. $\frac{51}{36} = \frac{17}{12}$ or $1\frac{5}{12}$ 7. $\frac{25}{42}$ 8. $\frac{29}{108}$

9. $\frac{59}{12}$ or $4\frac{11}{12}$ 10. $\frac{104}{15}$ or $6\frac{14}{15}$ 11. $\frac{51}{8}$ or $6\frac{3}{8}$

12. $\frac{409}{48}$ or $8\frac{25}{48}$ 13. $\frac{124}{15}$ or $8\frac{4}{15}$ 14. $\frac{57}{10}$ or $5\frac{7}{10}$

Developmental Arithmetic–A Computational Review

lesson ten

SUBTRACTION OF FRACTIONS

In the last lesson we reviewed what must be done when we add two fractions with different denominators. When we subtract two fractions, we again find that it works best to have a common denominator that is as small as possible. The rule we use is as follows:

> **Find the common denominator; change each fraction to a new fraction with the common denominator; and place the difference of the numerators over the common denominator.**

The Rule For Subtracting Fractions

Let's start by looking at the following example. Again, we will use the three step method we used for adding fractions. But in the last step we will subtract the numerators instead of adding.

$$\frac{2}{3} - \frac{5}{8}$$

Step 1.

×	Multiples of 3	Multiples of 8
1	3	8
2	6	16
3	9	24
4	12	32
5	15	
6	18	
7	21	
8	24	

Lesson Ten—Subtraction of Fractions

97

Subtracting Common Fractions With Different Denominators

Step 2. Our smallest common multiple is 24. Again, we use the basic principle of fractions to change each fraction.

$$\frac{2}{3} = \frac{2 \times 8}{3 \times 8} = \frac{16}{24} \qquad \frac{5}{8} = \frac{5 \times 3}{8 \times 3} = \frac{15}{24}$$

Step 3.

$$\begin{array}{r} \frac{2}{3} = \frac{16}{24} \\ -\ \frac{5}{8} = \frac{15}{24} \\ \hline \frac{1}{24} \end{array}$$

Most of the time subtracting fractions is as easy as adding fractions. However, sometimes subtracting mixed numbers presents special problems. For example, let's subtract $1\frac{1}{3}$ from $3\frac{1}{4}$.

Subtracting Mixed Numbers

Step 1. The least common multiple is 12.

Step 2. We change $\frac{1}{4}$ and $\frac{1}{3}$ to new fractions with 12 as the new denominator.

$$\frac{1}{4} = \frac{1 \times 3}{4 \times 3} = \frac{3}{12} \qquad \frac{1}{3} = \frac{1 \times 4}{3 \times 4} = \frac{4}{12}$$

Step 3. We can rewrite the original problem as

$$\begin{array}{r} 3\frac{1}{4} = 3\frac{3}{12} \\ -\ 1\frac{1}{3} = 1\frac{4}{12} \\ \hline \end{array}$$

98   Developmental Arithmetic–A Computational Review

We cannot subtract $\frac{4}{12}$ from $\frac{3}{12}$; so, we need to borrow 1 from the 3. Before we can use the 1 we borrowed, we need to change it to a fraction that can be added to the $\frac{3}{12}$. We change the 1 to $\frac{12}{12}$ and add it to the $\frac{3}{12}$, getting the fraction $\frac{15}{12}$. Remember to change the 3 to a 2 after doing the borrowing.

Borrowing With Mixed Numbers

$$\begin{array}{r} \overset{2}{\cancel{3}}\overset{15}{\frac{3}{12}} \\ -\ 1\frac{4}{12} \\ \hline 1\frac{11}{12} \end{array}$$

WARNING: A common error is made when 1 is borrowed in fraction problems. Remember, you are <u>not</u> borrowing a ten. The fraction $\frac{3}{12}$ does not become $\frac{13}{12}$. The correct result of the borrowing is $1\frac{3}{12}$ which becomes the improper fraction $\frac{15}{12}$. When you borrow, be careful!

We could do the above problem without borrowing and avoid the possible errors. All we need to do is first change each mixed number to an improper fraction as we did earlier with the addition of mixed numbers in the last lesson.

Using Improper Fractions To Avoid Borrowing

$$3\frac{1}{4} = \frac{13}{4} = \frac{13 \times 3}{4 \times 3} = \frac{39}{12}$$

$$1\frac{1}{3} = \frac{4}{3} = \frac{4 \times 4}{3 \times 4} = \frac{16}{12}$$

Lesson Ten–Subtraction of Fractions

Now, we subtract the improper fractions as follows:

$$\begin{array}{r} \dfrac{39}{12} \\ - \dfrac{16}{12} \\ \hline \dfrac{23}{12} = 1\dfrac{11}{12} \end{array}$$

Of course, if we do not have to borrow, we can subtract mixed numbers easily. For example:

$$4\dfrac{1}{2} - 2\dfrac{1}{3}$$

Step 1. The LCM is 6.

Subtracting Mixed Numbers Without Borrowing

Step 2. We change the fractions as follows:

$$\dfrac{1}{2} = \dfrac{3}{6} \qquad \dfrac{1}{3} = \dfrac{2}{6}$$

Step 3. Now we subtract.

$$\begin{array}{r} 4\dfrac{3}{6} \\ - \ 2\dfrac{2}{6} \\ \hline 2\dfrac{1}{6} \end{array}$$

REVIEW

- **Subtraction of Fractions** *See page 97*
  In order to subtract fractions, it is necessary for the fractions to have common denominators. We then subtract the numerators and place the answer over the common denominator.

- **Subtracting Mixed Numbers** *See page 98*
  When it is necessary to borrow one from the whole number during the subtraction process of mixed numbers, we first need to change the 1 to a fraction with the same denominator as the fraction in the mixed number.

- **Mixed Numbers Versus Improper Fractions** *See page 99*
  Mixed numbers can be changed to improper fractions to avoid borrowing when subtracting mixed numbers.

EXERCISES

Subtract the following fractions.

1. $\frac{3}{5} \quad \frac{12}{20}$
   $-\frac{1}{4} \quad \frac{5}{20}$
   _____
   $\frac{7}{20}$

2. $\frac{1}{2} \quad \frac{3}{6}$
   $-\frac{1}{3} \quad \frac{2}{6}$
   _____
   $\frac{1}{6}$

3. $\frac{5}{6}$
   $-\frac{1}{2}$
   _____
   $\frac{1}{3}$

4. $\frac{7}{12} \quad \frac{14}{24}$
   $-\frac{3}{8} \quad \frac{9}{24}$
   _____
   $\frac{5}{24}$

5. $\frac{7}{18} \quad \frac{56}{144}$
   $-\frac{3}{16} \quad \frac{27}{144}$
   _____
   $\frac{29}{144}$

6. $\frac{13}{20} \quad \frac{91}{140}$
   $-\frac{4}{35} \quad \frac{16}{140}$
   _____
   $\frac{75}{140}$

102   Developmental Arithmetic–A Computational Review

7. $\frac{5}{6} \quad 3\frac{15}{18}$
   $-\frac{7}{9} \quad 2\frac{14}{18}$
   _____
   $\frac{1}{18}$

8. $4\frac{7}{8}$
   $-2\frac{1}{4} \quad 2\frac{2}{8}$
   _____
   $2\frac{5}{8}$

9. $4\frac{2}{3} \quad 4\frac{8}{12}$
   $-1\frac{1}{4} \quad 3\frac{3}{12}$
   _____
   $3\frac{5}{12}$

10. $5\frac{1}{4} \quad \frac{5}{4} \quad 3\frac{15}{12}$
    $-2\frac{2}{3} \quad \quad 4\frac{8}{12}$
    _____
    $2\frac{7}{12}$

11. $5\frac{2}{3} \quad \frac{5}{3} \quad 4\frac{20}{12}$
    $-\frac{3}{4} \quad 3\frac{9}{12}$
    _____
    $4\frac{11}{12}$

12. $4 \quad \frac{3}{3}$
    $-2\frac{2}{3}$
    _____
    $1\frac{1}{3}$

Lesson Ten—Subtraction of Fractions

103

ANSWERS TO EXERCISES

1. $\dfrac{3}{5} = \dfrac{3\times 4}{5\times 4} = \dfrac{12}{20}$
    $-\dfrac{1}{4} = \dfrac{1\times 5}{4\times 5} = \dfrac{5}{20}$
    $\phantom{-\dfrac{1}{4} = \dfrac{1\times 5}{4\times 5} =}\dfrac{7}{20}$

2. $\dfrac{1}{2} = \dfrac{1\times 3}{2\times 3} = \dfrac{3}{6}$
    $-\dfrac{1}{3} = \dfrac{1\times 2}{3\times 2} = \dfrac{2}{6}$
    $\phantom{-\dfrac{1}{3} = \dfrac{1\times 2}{3\times 2} =}\dfrac{1}{6}$

3. $\dfrac{2}{6} = \dfrac{1}{3}$ 4. $\dfrac{5}{24}$ 5. $\dfrac{29}{144}$

6. $\dfrac{75}{140} = \dfrac{15}{28}$ 7. $\dfrac{1}{18}$ 8. $2\dfrac{5}{8}$

9. $3\dfrac{5}{12}$ 10. $2\dfrac{7}{12}$ 11. $4\dfrac{11}{12}$

12. $1\dfrac{1}{3}$

lesson eleven

THE LEAST COMMON DENOMINATOR

In the last two lessons we learned that we needed to find a common denominator to add and subtract fractions. We found that there are two ways to get the smallest common denominator. The first way was simply by inspection and a knowledge of multiplication facts. This method works well with small numbers.

We also learned about using lists of multiples as a second method for finding the smallest common denominator. We used the smallest multiple common to both denominators and used that number as the new denominator; calling it the **least common denominator** or LCD.

Least Common Denominator (LCD)

There is a third method that can be used to determine the least common denominator. We can use prime factoring to find the LCD.

The three steps to follow to use prime factoring to find the LCD:

Step 1. Use prime factoring to find the prime factors of each denominator.

Step 2. Write each denominator as a product of primes using exponents.

The Three Step Method for Finding the LCD

Step 3. Write a product of the primes found in either product of primes. Use each prime <u>only once</u>. Then write in the <u>largest exponent</u> used on that prime in the prime factoring of the denominators.

Lesson Eleven–The Least Common Denominator 105

We will use the three step method to find the least common denominator for the sum of the following two fractions:

$$\frac{5}{18} + \frac{7}{24}$$

Finding the LCD of 18 and 24

Step 1. Use prime factoring to find the prime factors of each denominator.

$$\begin{array}{r}1\\3\overline{)3}\\3\overline{)9}\\2\overline{)18}\end{array} \qquad \begin{array}{r}1\\3\overline{)3}\\2\overline{)6}\\2\overline{)12}\\2\overline{)24}\end{array}$$

Step 2. Write each denominator as a product of primes using exponents.

$$18 = 2^1 \times 3^2 \qquad\qquad 24 = 2^3 \times 3^1$$

Step 3. Write a product of the primes found in either product of primes. Use each prime <u>only once</u>.

$$2 \times 3.$$

Then write in the <u>largest exponent</u> used on that prime in the prime factoring of the denominators.

When we compare 2^1 and 2^3, we note that the largest exponent used on the prime 2 was a 3.

When we compare 3^2 and 3^1, we note that the largest exponent used on the prime 3 was a 2.

We now write in the largest exponent used on each prime to get the LCD as follows.

$$LCD = 2^3 \times 3^2 = 72$$

We then change each fraction so it has the new denominator and add the fractions as follows:

$$\frac{5 \times 4}{18 \times 4} = \frac{20}{72} \qquad \frac{7 \times 3}{24 \times 3} = \frac{21}{72}$$

$$\frac{20}{72} + \frac{21}{72} = \frac{20 + 21}{72} = \frac{41}{72}$$

We will look at a second example involving a subtraction:

$$\frac{9}{20} - \frac{7}{30}$$

Step 1. Use prime factoring to find the prime factors of each denominator.

Finding the LCD of 20 and 30

$$\begin{array}{r} 1 \\ 5\overline{)5} \\ 2\overline{)10} \\ 2\overline{)20} \end{array} \qquad \begin{array}{r} 1 \\ 5\overline{)5} \\ 3\overline{)15} \\ 2\overline{)30} \end{array}$$

Step 2. Write each denominator as a product of primes using exponents.

$$20 = 2^2 \times 5^1 \qquad 30 = 2^1 \times 3^1 \times 5^1$$

Step 3. Write a product of the primes found in either product of primes. Use each prime <u>only once</u>. Then write in the <u>largest exponent</u> used on that prime in the prime factoring of the denominators.

$$LCD = 2^2 \times 3^1 \times 5^1 = 60$$

$$\frac{9 \times 3}{20 \times 3} = \frac{27}{60} \qquad \frac{7 \times 2}{30 \times 2} = \frac{14}{60}$$

$$\frac{27}{60} - \frac{14}{60} = \frac{27 - 14}{60} = \frac{13}{60}$$

Lesson Eleven–The Least Common Denominator

This time we will find the LCD for the denominators of the fractions $\frac{7}{10}$ and $\frac{5}{21}$. We will not add or subtract the fractions. We are going to learn something more about finding LCDs.

Step 1. Use prime factoring to find the prime factors of each denominator.

Finding the LCD of 10 and 21

$$\begin{array}{r}1\\5\overline{)5}\\2\overline{)10}\end{array} \qquad \begin{array}{r}1\\7\overline{)7}\\3\overline{)21}\end{array}$$

Step 2. Write each denominator as a product of primes using exponents.

$$10 = 2^1 \times 5^1 \qquad 21 = 3^1 \times 7^1$$

Step 3. Write a product of the primes found in either product of primes. Use each prime <u>only once</u>; with the <u>largest exponent</u> used on that prime in the prime factoring of the denominators.

$$LCD = 2^1 \times 3^1 \times 5^1 \times 7^1 = 210$$

Notice that we do not have any primes that are in both products. That is, we do not have any common prime factors. When this happens, we say that the two denominators are **relatively prime.**

Relatively Prime Denominators

An easier way of deciding if two denominators are relatively prime involves a simple division test. The two denominators are relatively prime if 1 is the only number that divides both of them; or more simply, 1 is the only common divisor of relatively prime numbers. When two denominators are relatively prime, we simply multiply the two denominators to get the LCD.

108 Developmental Arithmetic–A Computational Review

> **If 1 is the only number that divides both denominators, the two denominators are relatively prime.**

What To Do If Relatively Prime Denominators

Returning to our example involving the fractions $\frac{7}{10}$ and $\frac{5}{21}$, the only common divisor of both 10 and 21 is 1. Since the denominators are relatively prime, the LCD can be found by multiplying the two denominators together.

$$LCD = 10 \times 21 = 210$$

> **Remember: If the denominators are relatively prime, the product of the denominators is the LCD.**

Lesson Eleven–The Least Common Denominator

REVIEW

See page 105

- **Least Common Denominator Using Prime Factors**
  The three step process can be used to find the LCD:

  Step 1. Use prime factoring to find the prime factors of each denominator.

  Step 2. Write each denominator as a product of primes using exponents.

  Step 3. Write a product of the primes found in either product of primes. Use each prime only once. Then write in the largest exponent used on that prime in the prime factoring of the denominators.

See page 108

- **Relatively Prime Denominators**
  Two denominators are relatively prime if the numbers have 1 as the only common divisor.
  If the denominators are relatively prime, the LCD is found by multiplying the two denominators.

EXERCISES

Use the three step process to find the LCD for the denominators; then change each fraction to a new fraction with the LCD. Finally add or subtract the fractions as indicated.

1. $\dfrac{5}{24} + \dfrac{7}{36}$

2. $\dfrac{4}{45} + \dfrac{5}{27}$

3. $\dfrac{8}{35} - \dfrac{2}{21}$

4. $\dfrac{7}{16} - \dfrac{5}{21}$

Lesson Eleven—The Least Common Denominator

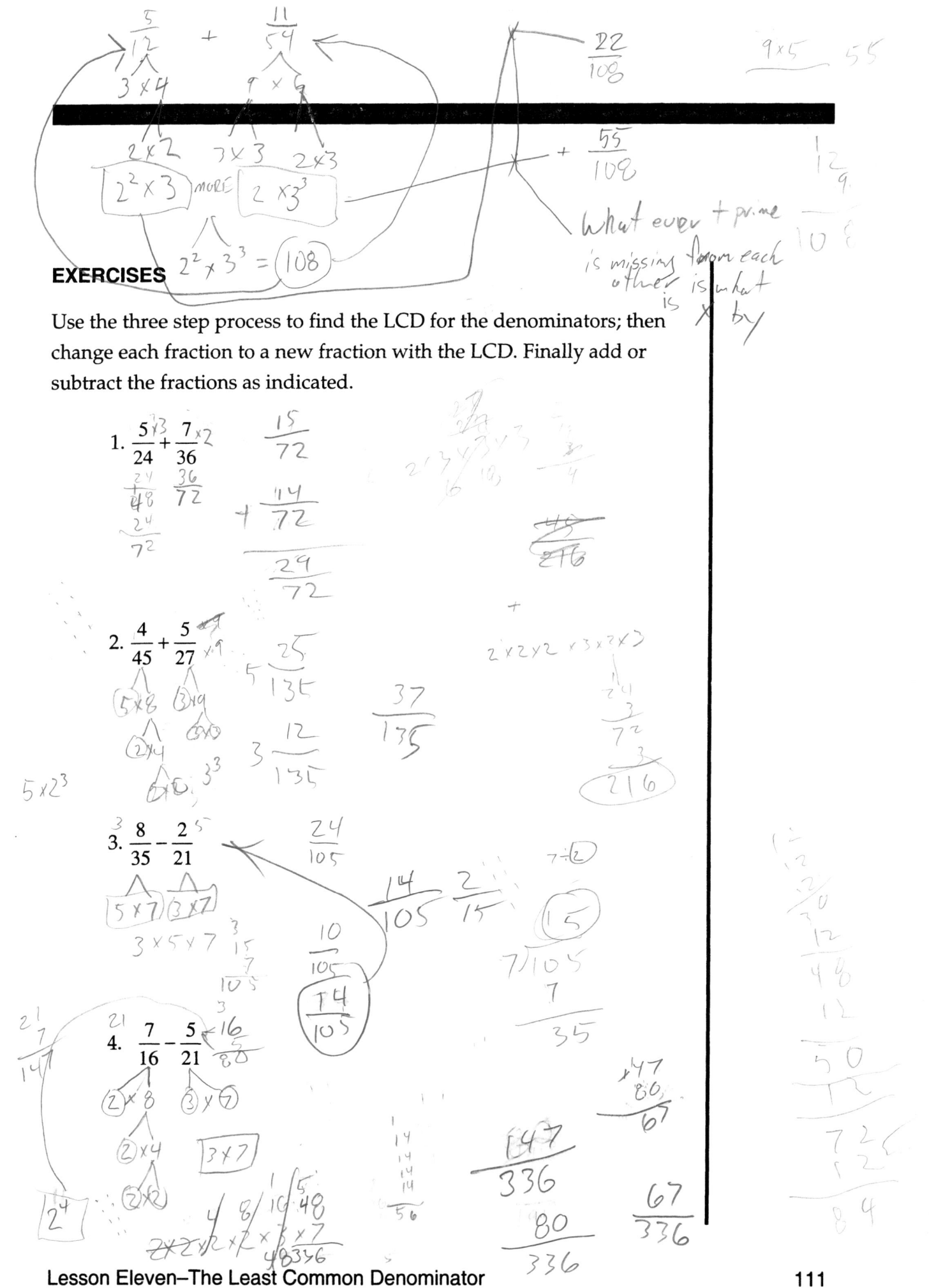

5. $\dfrac{7}{18} + \dfrac{5}{12}$

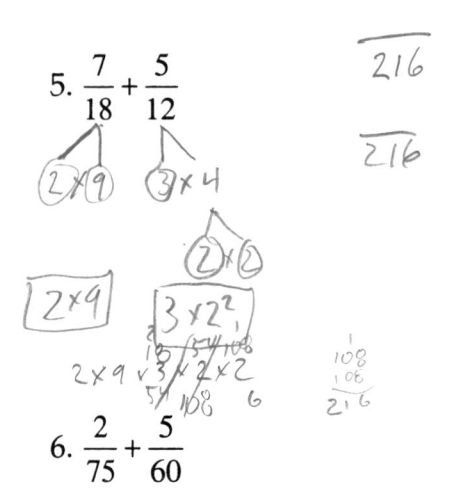

6. $\dfrac{2}{75} + \dfrac{5}{60}$

7. $\dfrac{7}{90} - \dfrac{5}{72}$

8. $\dfrac{11}{36} - \dfrac{3}{25}$

ANSWERS TO EXERCISES

1. $LCD = 2^3 \times 3^2 = 2 \times 2 \times 2 \times 3 \times 3 = 72$

$$\frac{5}{24} = \frac{15}{72} \qquad \frac{7}{36} = \frac{14}{72}$$

$$\frac{15}{72} + \frac{14}{72} = \frac{29}{72}$$

2. $LCD = 3^3 \times 5^1 = 3 \times 3 \times 3 \times 5 = 135$

$$\frac{4}{45} = \frac{12}{135} \qquad \frac{5}{27} = \frac{25}{135}$$

$$\frac{12}{135} + \frac{25}{135} = \frac{37}{135}$$

3. $LCD = 3^1 \times 5^1 \times 7^1 = 3 \times 5 \times 7 = 105$

$$\frac{8}{35} = \frac{24}{105} \qquad \frac{2}{21} = \frac{10}{105}$$

$$\frac{24}{105} - \frac{10}{105} = \frac{14}{105} = \frac{2}{15}$$

4. $LCD = 2^4 \times 3^1 \times 7^1 = 2 \times 2 \times 2 \times 2 \times 3 \times 7 = 336$

$$\frac{7}{16} = \frac{147}{336} \qquad \frac{5}{21} = \frac{80}{336}$$

$$\frac{147}{336} - \frac{80}{336} = \frac{67}{336}$$

Lesson Eleven–The Least Common Denominator

5. LCD $= 2^2 \times 3^2 = 2 \times 2 \times 3 \times 3 = 36$

$$\frac{7}{18} = \frac{14}{36} \qquad \frac{5}{12} = \frac{15}{36}$$

$$\frac{14}{36} + \frac{15}{36} = \frac{29}{36}$$

6. LCD $= 2^2 \times 3^1 \times 5^2 = 2 \times 2 \times 3 \times 5 \times 5 = 300$

$$\frac{2}{75} = \frac{8}{300} \qquad \frac{5}{60} = \frac{25}{300}$$

$$\frac{8}{300} + \frac{25}{300} = \frac{33}{300} = \frac{11}{100}$$

7. LCD $= 2^3 \times 3^2 \times 5^1 = 2 \times 2 \times 2 \times 3 \times 3 \times 5 = 360$

$$\frac{7}{90} = \frac{28}{360} \qquad \frac{5}{72} = \frac{25}{360}$$

$$\frac{28}{360} - \frac{25}{360} = \frac{3}{360} = \frac{1}{120}$$

8. LCD $= 2^2 \times 3^2 \times 5^2 = 2 \times 2 \times 3 \times 3 \times 5 \times 5 = 900$

$$\frac{11}{36} = \frac{275}{900} \qquad \frac{3}{25} = \frac{108}{900}$$

$$\frac{275}{900} - \frac{108}{900} = \frac{167}{900}$$

lesson twelve

MULTIPLICATION OF FRACTIONS

Multiplication of common fractions is done using a simple rule:

The Simple Rule For Multiplying Fractions

> **Place the product of the numerators over the product of the denominators. If necessary, reduce the answer.**

For example:

$$\frac{1}{2} \times \frac{2}{3} = \frac{1 \times 2}{2 \times 3} = \frac{2}{6} = \frac{1}{3}$$

We can show the multiplication of $\frac{1}{2}$ and $\frac{2}{3}$ geometrically.

The diagram for $\frac{1}{2}$ will be:

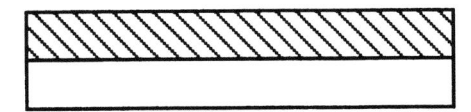

A Geometric Interpretation Of Multiplying Fractions

The diagram for $\frac{2}{3}$ will be:

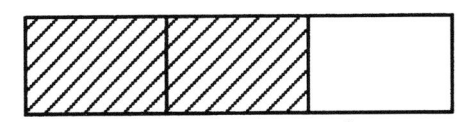

Now, think of lifting the diagram for $\frac{1}{2}$ and placing it on top of the diagram for $\frac{2}{3}$. The diagram below shows the result.

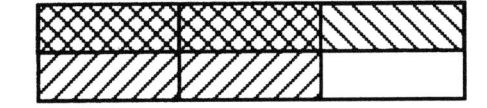

Lesson Twelve—Multiplication of Fractions

There are 6 parts and 2 of the 6 parts are cross-hatched ⊠. The $\frac{2}{6}$ ($\frac{1}{3}$ when reduced) is the same answer we obtained when we multiplied using the simple rule.

You may be bothered that $\frac{1}{2} \times \frac{2}{3} = \frac{1}{3}$ gives a smaller number than either of the factors $\frac{1}{2}$ and $\frac{2}{3}$. Although we use the process of multiplication to find $\frac{1}{2} \times \frac{2}{3}$, we really are asking, "What is $\frac{1}{2}$ of $\frac{2}{3}$?" Clearly, $\frac{1}{2}$ of something must be smaller than the original number. When whole numbers are multiplied, the answers are always larger. But when we multiply proper common fractions the product will always be smaller than either factor.

The simple rule we used for multiplying $\frac{1}{2} \times \frac{2}{3}$ often yields large numbers in both the numerator and the denominator. Large numbers are usually difficult to reduce. We can avoid the difficulty by using a

Cancelling

method you may remember as **cancelling**; it is sometimes referred to as cross cancelling when all of the factors are themselves reduced.

It is important to remember that cancelling requires that we divide the same number into both the numerator and the denominator of the fraction. For example, if we divide the numerator by 2, we must also divide the denominator by 2. Actually when we cancel a fraction,

Basic Principle Of Fractions

we are again using the **basic principle of fractions**.

The example below illustrates the cancelling method. Let's look at each step. We start by dividing 25 and 15 by 5. This is a **"first level"** cancelling.

$$\frac{18}{2\cancel{5}} \times \frac{\cancel{15}^{3}}{12}$$
$$_{5}$$

Now, we can divide both 18 and 12 by 6. This gives one more "first level" cancelling that involves finding the largest number that will divide both the numerator and the denominator.

First Level Cancelling

$$\frac{\cancel{18}^{3}}{2\cancel{5}} \times \frac{\cancel{15}^{3}}{\cancel{12}}$$
$$_{5} _{2}$$

We now multiply the numbers that remain in the numerator and denominator to get our answer of $\frac{3 \times 3}{5 \times 2}$ or $\frac{9}{10}$.

We could have done the problem in several other ways. For example rather than cross cancelling, we could have started by reducing the second factor as a first step. Or we could start by cancelling with smaller numbers like 2s and 3s. In each case the cancelling would look different, but we would get the same answer at the end. Four other versions of the cancellations are shown below:

$$\frac{\cancel{18}^{9}}{2\cancel{5}} \times \frac{\cancel{15}^{5}}{\cancel{12}} \qquad\qquad \frac{\cancel{18}^{9}}{2\cancel{5}} \times \frac{\cancel{15}^{5}}{\cancel{12}}$$
$$_{5} _{\substack{4\\2}} _{5} _{\substack{6\\2}}$$

$$\frac{\cancel{\cancel{18}^{6}}^{3}}{2\cancel{5}} \times \frac{\cancel{15}^{3}}{\cancel{12}} \qquad\qquad \frac{\cancel{\cancel{18}^{9}}^{3}}{2\cancel{5}} \times \frac{\cancel{15}^{3}}{\cancel{12}}$$
$$_{5} _{\substack{4\\2}} _{5} _{\substack{6\\2}}$$

In each of the four solutions a "**second level**" of cancelling was used because we did not use the largest divisor possible. A "second level" cancelling is done when we end up having to cancel a number we obtained from an earlier first level cancelling. We often make errors in "second level" or even "higher level" cancelling because the cancelling

Second Level Cancelling

Lesson Twelve–Multiplication of Fractions

Using Prime Factors To Cancel In Multiplication

starts to look cluttered. If we get just a little bit sloppy in our work, it is very easy to make careless errors. Sometimes we just do not see the largest divisor that can be used to cancel. And in some instances it is impossible to avoid higher level cancelling. At such times, the prime factoring method may result in fewer errors. We used this method earlier in Lesson 7 to reduce fractions.

The first step in using the method of prime factors is to simply write the product as one fraction as follows:

$$\frac{18 \times 15}{25 \times 12}$$

But, we don't actually do the multiplication. We write the numbers in the numerator and denominator as products of primes:

$$18 = 2 \times 3 \times 3 \qquad 15 = 3 \times 5$$
$$25 = 5 \times 5 \qquad 12 = 2 \times 2 \times 3$$

Next, we rewrite the product in factored form as follows:

$$\frac{18 \times 15}{25 \times 12} = \frac{(2 \times 3 \times 3) \times (3 \times 5)}{(5 \times 5) \times (2 \times 2 \times 3)}$$

We can now cancel or reduce any common factors in the numerator and denominator to 1's. Please remember, if you cancel a number in the numerator, you must cancel that same number in the denominator.

$$\frac{\overset{1}{\cancel{2}} \times \overset{1}{\cancel{3}} \times 3 \times 3 \times \overset{1}{\cancel{5}}}{\underset{1}{\cancel{5}} \times 5 \times \underset{1}{\cancel{2}} \times 2 \times \underset{1}{\cancel{3}}}$$

Finally, multiply what is left in the numerator and write the product over the product of the denominator:

$$\frac{3 \times 3}{5 \times 2} = \frac{9}{10}$$

118   Developmental Arithmetic–A Computational Review

We now have three approaches that we can use to multiply.

> 1. **Multiply and reduce.**
> 2. **First level cancelling (or second level, if you can be accurate).**
> 3. **Prime factoring.**

Three Approaches That Can Be Used To Multiply Fractions

You will need to use your common sense when choosing the "right" method. If the problems are easy, either the "simple rule" or the cancelling method works well. Using the "prime factors" method is usually preferred if the numbers get very large. Perhaps you will get a better idea of which method to use as we go through more examples.

Let's look at the product of a simple fraction and a mixed number.

$$1\frac{2}{3} \times \frac{3}{4}$$

Simple Fractions Times Mixed Numbers

Rewrite the mixed number as an improper fraction:

$$1\frac{2}{3} = \frac{3+2}{3} = \frac{5}{3}$$

Using the "simple rule," we multiply as follows:

$$\frac{5}{3} \times \frac{3}{4} = \frac{5 \times 3}{3 \times 4} = \frac{15}{12}$$

We can divide both 15 and 12 by 3 and the fraction is then reduced to:

$$\frac{15 \div 3}{12 \div 3} = \frac{5}{4}$$

You probably saw a shortcut using "first level" cancelling.

$$\frac{5 \times \overset{1}{\cancel{3}}}{\underset{1}{\cancel{3}} \times 4} = \frac{5}{4}$$

Lesson Twelve—Multiplication of Fractions

It would make very little sense to use prime factoring in the problem. Nothing is gained as shown below:

$$\frac{5 \times \overset{1}{\cancel{3}}}{\underset{1}{\cancel{3}} \times 2 \times 2} = \frac{5}{4}$$

Multiplying More Than Two Fractions Together

Sometimes we must multiply more than two fractions together. For example:

$$\frac{2}{5} \times \frac{10}{21} \times \frac{9}{16}$$

We can do the problem as two separate problems using "first level" cancelling as follows:

$$\frac{2}{\underset{1}{\cancel{5}}} \times \frac{\overset{2}{\cancel{10}}}{21} = \frac{4}{21}$$

$$\frac{\overset{1}{\cancel{4}}}{\underset{7}{\cancel{21}}} \times \frac{\overset{3}{\cancel{9}}}{\underset{4}{\cancel{16}}} = \frac{3}{28}$$

We could also do the above problem trying to use "first level" cancelling as one problem, but we end up having to do a "second level" cancelling:

$$\frac{\overset{1}{\cancel{2}}}{\underset{1}{\cancel{5}}} \times \frac{\overset{\overset{1}{\cancel{2}}}{\cancel{10}}}{\underset{7}{\cancel{21}}} \times \frac{\overset{3}{\cancel{9}}}{\underset{\underset{4}{8}}{\cancel{16}}} = \frac{3}{28}$$

Or, we can avoid the "second level" cancelling by using the prime factoring method. We multiply the three fractions together as follows:

$$\frac{2}{5} \times \frac{10}{21} \times \frac{9}{16}$$

120   Developmental Arithmetic–A Computational Review

$$\frac{\overset{1}{\cancel{2}}\times\overset{1}{\cancel{2}}\times\overset{1}{\cancel{5}}\times\overset{1}{\cancel{3}}\times 3}{\underset{1}{\cancel{5}}\times\underset{1}{\cancel{3}}\times 7\times\underset{1}{\cancel{2}}\times\underset{1}{\cancel{2}}\times 2\times 2} = \frac{3}{28}$$

We can also multiply mixed numbers together. It is necessary to first rewrite them as improper fractions:

Multiplying Mixed Numbers

$$2\frac{1}{5}\times 3\frac{1}{8}$$

$$\frac{11}{5}\times\frac{25}{8} = \frac{11\times\overset{5}{\cancel{25}}}{\underset{1}{\cancel{5}}\times 8} = \frac{55}{8} \text{ or } 6\frac{7}{8}$$

Finally, we can multiply whole numbers times simple fractions or mixed numbers. The whole number is first changed to a common fraction with a denominator of 1.

Whole Numbers Times Simple Fractions Or Mixed Numbers

$$\frac{2}{3}\times 4\times 3\frac{1}{2}$$

$$\frac{2}{3}\times\frac{4}{1}\times\frac{7}{2} = \frac{\overset{1}{\cancel{2}}\times 4\times 7}{3\times 1\times\underset{1}{\cancel{2}}} = \frac{28}{3} \text{ or } 9\frac{1}{3}$$

Notice that in the above two examples we used "first level" cancelling. In both instances it would not have been profitable to use the "prime factors" approach. However, you may find it worthwhile to use the "prime factors" approach in some of the problems in the exercises for this lesson.

Lesson Twelve—Multiplication of Fractions

REVIEW

See page 115

- **To Multiply Common Fractions**
  Place the product of the numerators over the product of the denominators. If necessary, the result should be reduced.

See page 116

- **Cancelling**
  A common divisor is divided into both the numerator and the denominator to simplify the multiplication of fractions. It is best to use "first level" cancelling.

See page 118

- **Multiplying Fractions Using Prime Factoring**
  The "prime factoring" method is sometimes used to avoid careless errors in "higher level" cancelling. After factoring each number into primes, common factors in the numerator and the denominator are cancelled. All factors left in the numerator and denominator are then multiplied to get the final product.

See page 119

- **To Multiply Mixed Numbers With Common Fractions**
  It is necessary to first change the mixed number to an improper fraction.

See page 120

- **Multiplying More Than Two Fractions Together**
  Multiply the fractions two at a time or all at once or by using "prime factoring". Then use cancelling.

See page 121

- **To Multiply Whole Numbers With Fractions**
  Change the whole number to a common fraction by putting the whole number over a denominator of 1.

EXERCISES

Find the following products. All answers should be reduced.

1. $\frac{1}{2} \times \frac{1}{3} = \frac{1}{6}$

2. $\frac{6}{7} \times \frac{14}{15} = \frac{4}{5}$

3. $\frac{3}{8} \times 2\frac{1}{8} = \frac{3}{4}$

4. $\frac{25}{28} \times \frac{7}{10} = \frac{5}{8}$

5. $\frac{32}{63} \times \frac{21}{40}$

6. $4\frac{2}{3} \times \frac{2}{3} = \frac{28}{9} = 3\frac{1}{9}$

7. $5\frac{2}{7} \times 3\frac{1}{2} = \frac{35}{2} = 17\frac{1}{2}$

8. $\frac{2}{3} \times \frac{5}{8} \times \frac{6}{7} = \frac{5}{14}$

9. $2 \times 1\frac{7}{8} \times \frac{1}{3} = \frac{5}{4} = 1\frac{1}{4}$

10. $2\frac{1}{6} \times 3\frac{3}{5} \times 1\frac{2}{3} = 13$

Lesson Twelve—Multiplication of Fractions

ANSWERS TO EXERCISES

1. $\dfrac{1}{2} \times \dfrac{1}{3} = \dfrac{1}{6}$

2. $\dfrac{\overset{2}{\cancel{6}}}{\underset{1}{\cancel{7}}} \times \dfrac{\overset{2}{\cancel{14}}}{\underset{5}{\cancel{15}}} = \dfrac{4}{5}$

3. $\dfrac{3}{\underset{4}{\cancel{8}}} \times \dfrac{\overset{1}{\cancel{2}}}{1} = \dfrac{3}{4}$

4. $\dfrac{\overset{5}{\cancel{25}}}{\underset{4}{\cancel{28}}} \times \dfrac{\overset{1}{\cancel{7}}}{\underset{2}{\cancel{10}}} = \dfrac{5}{8}$

5. $\dfrac{\overset{4}{\cancel{32}}}{\underset{3}{\cancel{63}}} \times \dfrac{\overset{1}{\cancel{21}}}{\underset{5}{\cancel{40}}} = \dfrac{4}{15}$

6. $\dfrac{14}{3} \times \dfrac{2}{3} = \dfrac{28}{9}$ or $3\dfrac{1}{9}$

7. $\dfrac{37}{\underset{1}{\cancel{7}}} \times \dfrac{\cancel{7}}{2} = \dfrac{37}{2}$ or $18\dfrac{1}{2}$

8. $\dfrac{\overset{1}{\cancel{2}}}{\underset{1}{\cancel{3}}} \times \dfrac{5}{\underset{2}{\cancel{8}}_4} \times \dfrac{\overset{\frac{1}{2}}{\cancel{6}}}{7} = \dfrac{5}{14}$

9. $\dfrac{\overset{1}{\cancel{2}}}{1} \times \dfrac{\overset{5}{\cancel{15}}}{\underset{4}{\cancel{8}}} \times \dfrac{1}{\underset{1}{\cancel{3}}} = \dfrac{5}{4}$ or $1\dfrac{1}{4}$

10. $\dfrac{13}{\underset{1}{\cancel{6}}} \times \dfrac{\overset{\frac{1}{3}}{\cancel{18}}}{\underset{1}{\cancel{5}}} \times \dfrac{\overset{1}{\cancel{5}}}{\underset{1}{\cancel{3}}} = \dfrac{13}{1}$

lesson thirteen

DIVISION OF FRACTIONS

If you are familiar with the rules and methods for multiplying fractions, division of fractions requires only one new rule.

> **To divide one fraction by another, invert the divisor and multiply. The divisor is the fraction after the ÷ sign.**

The Rule For Dividing Simple Fractions

Let's look at an example and maybe you will remember the rule.

$$\frac{2}{3} \div \frac{2}{5} = \frac{2}{3} \times \frac{5}{2}$$

The <u>second</u> fraction called the divisor, $\frac{2}{5}$, was <u>inverted</u> and the division sign was changed to a multiplication sign. Note, $\frac{5}{2}$ is called the **reciprocal** or the **multiplicative inverse** of $\frac{2}{5}$. At this point we can use "first level" cancelling to get the answer. One word of caution is necessary. **Never cancel before inverting the divisor** and rewriting the division as a multiplication. Only after the division problem has been written as a multiplication can we use the cancelling shortcut.

Reciprocal Or Multiplicative Inverse

Never Cancel Before Inverting The Divisor

$$\frac{2}{3} \div \frac{2}{5} = \frac{\overset{1}{\cancel{2}}}{3} \times \frac{5}{\underset{1}{\cancel{2}}} = \frac{5}{3} \text{ or } 1\frac{2}{3}$$

Lesson Thirteen—Division of Fractions

Rewriting An Improper Fraction Is Optional

Remember, writing $\frac{5}{3}$ as $1\frac{2}{3}$ is optional. We can leave the answer as an improper fraction. It is OK; we do not have to rewrite $\frac{5}{3}$ as a mixed number.

Now let's see why we invert the fraction after the division symbol. We need to remember that a fraction is just another way of writing a division problem. So we can write the division as follows with what we call a **complex fraction**:

Complex Fraction

$$\frac{2}{3} \div \frac{2}{5} = \frac{\frac{2}{3}}{\frac{2}{5}}$$

Next, we apply the **basic principle of fractions** to the complex fraction. We multiply both the numerator (top) and the denominator (bottom) of the complex fraction by $\frac{5}{2}$.

$$\frac{\frac{2}{3} \times \frac{5}{2}}{\frac{2}{5} \times \frac{5}{2}}$$

Notice that the multiplication of the bottom fractions results in $\frac{10}{10}$ or 1.

$$\frac{\frac{2}{3} \times \frac{5}{2}}{\frac{2}{5} \times \frac{5}{2}} = \frac{\frac{2}{3} \times \frac{5}{2}}{1} = \frac{\overset{1}{\cancel{2}}}{3} \times \frac{5}{\underset{1}{\cancel{2}}} = \frac{5}{3}$$

126   Developmental Arithmetic–A Computational Review

Hence, we have **the rule for dividing fractions that tells us to invert the divisor and multiply.** When students review division of fractions, most of them remember that they are supposed to invert and multiply. They just have trouble remembering which fraction gets inverted. Perhaps the following example will help remind us of which part of the division problem is inverted.

Rule For Dividing Fractions

Suppose we have four children who want to share two apples. If we give each child $\frac{1}{2}$ of an apple, we find that the 2 apples divided into $\frac{1}{2}$'s will give us 4 pieces of apple. We are simply stating the division problem 2 divided by $\frac{1}{2}$. In order to get the correct answer of 4 we must invert the $\frac{1}{2}$ and multiply. If we decided incorrectly to invert the first fraction, $\frac{2}{1}$, we would get a wrong answer of $\frac{1}{4}$ when we invert and multiply. Hence, the second fraction $\frac{1}{2}$ is inverted as follows:

$$\frac{2}{1} \div \frac{1}{2} = \frac{2}{1} \times \frac{2}{1} = \frac{4}{1}$$

As in the multiplication of fractions, **when we divide mixed numbers, we must first change the mixed numbers to improper fractions** as shown in the example below:

Dividing Mixed Numbers

$$1\frac{1}{2} \div 2\frac{1}{3} = \frac{3}{2} \div \frac{7}{3}$$

Lesson Thirteen–Division of Fractions   127

When we see the two 3s it is very tempting to use "first level" cancelling, but do not try to cancel yet. We must first invert the divisor and rewrite the division as a multiplication before we cancel. In this case after we invert, it is impossible to cancel anything. So we just multiply the numbers in the numerator and denominator to get

$$\frac{3}{2} \div \frac{7}{3} = \frac{3}{2} \times \frac{3}{7} = \frac{3 \times 3}{2 \times 7} = \frac{9}{14}$$

Order of Operations

Sometimes we must do problems that involve both multiplication and division of fractions. We will follow the usual **order of operations** that tells us to do multiplications and divisions first in the order we see them as we read the problem from left to right. The following example illustrates how we do such a problem.

$$\frac{3}{4} \div \frac{1}{2} \times \frac{2}{3}$$

Since the division comes first, we invert the $\frac{1}{2}$ (but not the $\frac{2}{3}$) and change the division to a multiplication. We then can use the method we used in the last lesson to multiply two or more fractions together.

$$\frac{\overset{1}{\cancel{3}}}{\underset{2}{\underset{1}{\cancel{4}}}} \times \frac{\overset{1}{\cancel{2}}}{1} \times \frac{\overset{1}{\cancel{2}}}{\underset{1}{\cancel{3}}} = \frac{1}{1} = 1$$

Notice that even though all the factors canceled, we were left with 1s on the top and bottom of the fraction. There is often a temptation to leave out the 1s when we do the cancelling and think that zero is the answer at this point in the problem. Instead we need to write in the 1s and then multiply them together to get the correct answer of 1.

Write In The 1s When Cancelling

We will look at one more example using the division rule to show how prime factoring can again be used to avoid careless errors that might occur due to "higher level" cancelling. Remember that we must first invert the divisor and then multiply.

$$\frac{15}{27} \div \frac{10}{21} = \frac{15}{27} \times \frac{21}{10} = \frac{\overset{1}{\cancel{15}} \times \overset{7}{\cancel{21}}}{\underset{9}{\cancel{27}} \times \underset{2}{\cancel{10}}} = \frac{7}{6} \text{ or } 1\frac{1}{6}$$

We may find it better to do the problem using the "prime factoring" method instead of the "second level" cancelling we did above.

$$\frac{15}{27} \div \frac{10}{21} = \frac{15}{27} \times \frac{21}{10} = \frac{\overset{1}{\cancel{3}} \times \overset{1}{\cancel{5}} \times \overset{1}{\cancel{3}} \times 7}{\underset{1}{\cancel{3}} \times \underset{1}{\cancel{3}} \times 3 \times 2 \times \underset{1}{\cancel{5}}} = \frac{7}{6} \text{ or } 1\frac{1}{6}$$

Lesson Thirteen–Division of Fractions

REVIEW

See page 125

- **Dividing Fractions**
  To divide one fraction by another fraction invert the divisor and multiply. Reduce the answer if necessary or use cancelling when doing the multiplication. Never cancel before inverting.

See page 127

- **To Divide Mixed Numbers and Fractions**
  It is necessary to first change the mixed number to an improper fraction, then invert the divisor, and multiply.

See page 128

- **Order of Operations**
  Do divisions and multiplications in the order given, going from left to right as we read the problem.

See page 129

- **Dividing Fractions Using Prime Factoring**
  The "prime factoring" method is sometimes used to avoid careless errors in "higher level" cancelling.

EXERCISES

Find the following quotients. All answers should be reduced.

1. $\dfrac{1}{3} \div \dfrac{2}{5}$

6. $\dfrac{1}{2} \div 2$

2. $\dfrac{3}{4} \div \dfrac{2}{3}$

7. $\dfrac{45}{77} \div \dfrac{27}{35}$

3. $\dfrac{8}{12} \div \dfrac{2}{3}$

8. $\dfrac{6}{35} \div 1\dfrac{1}{21}$

4. $\dfrac{4}{9} \div \dfrac{16}{15}$

9. $\dfrac{3}{5} \div \dfrac{9}{4} \times \dfrac{15}{16}$

5. $1\dfrac{1}{3} \div 2\dfrac{3}{8}$

10. $\dfrac{28}{36} \times \dfrac{25}{54} \div \dfrac{75}{42}$

Lesson Thirteen—Division of Fractions

ANSWERS TO EXERCISES

1. $\dfrac{1}{3} \times \dfrac{5}{2} = \dfrac{5}{6}$

6. $\dfrac{1}{2} \times \dfrac{1}{2} = \dfrac{1}{4}$

2. $\dfrac{3}{4} \times \dfrac{3}{2} = \dfrac{9}{8}$ or $1\dfrac{1}{8}$

7. $\dfrac{\overset{5}{\cancel{45}}}{\underset{11}{\cancel{77}}} \times \dfrac{\overset{5}{\cancel{35}}}{\underset{3}{\cancel{27}}} = \dfrac{25}{33}$

3. $\dfrac{\overset{\overset{1}{\cancel{4}}}{\cancel{8}}}{\underset{\underset{1}{\cancel{4}}}{\cancel{12}}} \times \dfrac{\overset{1}{\cancel{3}}}{\underset{1}{\cancel{2}}} = 1$

8. $\dfrac{\overset{3}{\cancel{6}}}{\underset{5}{\cancel{35}}} \times \dfrac{\overset{3}{\cancel{21}}}{\underset{11}{\cancel{22}}} = \dfrac{9}{55}$

4. $\dfrac{\overset{1}{\cancel{4}}}{\underset{3}{\cancel{9}}} \times \dfrac{\overset{5}{\cancel{15}}}{\underset{4}{\cancel{16}}} = \dfrac{5}{12}$

9. $\dfrac{\overset{1}{\cancel{3}}}{\underset{1}{\cancel{5}}} \times \dfrac{\overset{1}{\cancel{4}}}{\underset{3}{\cancel{9}}} \times \dfrac{\overset{\overset{1}{\cancel{5}}}{\cancel{15}}}{\underset{4}{\cancel{16}}} = \dfrac{1}{4}$

5. $\dfrac{4}{3} \times \dfrac{8}{19} = \dfrac{32}{57}$

10. $\dfrac{\overset{\overset{7}{\cancel{14}}}{\cancel{28}}}{\underset{\underset{3}{\cancel{6}}}{\cancel{36}}} \times \dfrac{\overset{1}{\cancel{25}}}{\underset{27}{\cancel{54}}} \times \dfrac{\overset{7}{\cancel{42}}}{\underset{3}{\cancel{75}}} = \dfrac{49}{243}$

lesson fourteen
DECIMAL FRACTIONS

A **decimal fraction** is the result of doing the division indicated by the common fraction. As was mentioned in Lesson 7, one way to think of a common fraction is as a division. For example, $\frac{1}{2}$ means $2\overline{)1}$. When we do the division, we get a decimal fraction.

Decimal Fractions

In order to write $\frac{1}{2}$ as a decimal fraction, do the following:

1. Place a **decimal point** after the 1 and add zero.

$$2\overline{)1}$$

Changing From A Common Fraction To A Decimal Fraction

2. Place a decimal point in the answer space, directly above the decimal point in 1.0

$$2\overline{)1.0}^{\,.}$$

3. Now divide and do not worry any further about the decimal point. But do be careful to write the digits in the correct position.

$$\begin{array}{r} 0.5 \\ 2\overline{)1.0} \\ \underline{1.0} \\ 0 \end{array}$$

4. We are done and we write the answer using positional notation as 0.5 or just .5

Lesson Fourteen–Decimal Fractions 133

Sometimes the decimal fractions we get from a common fraction do not end so quickly. For example, $\frac{1}{8}$ as a division gives .125 in positional notation. The division is shown below.

$$\begin{array}{r} .125 \\ 8\overline{)1.000} \\ \underline{8} \\ 20 \\ \underline{16} \\ 40 \\ \underline{40} \\ 0 \end{array}$$

This time we needed to add three zeros after the decimal point. We add a zero each time we want to get another digit in the answer until we are done.

Terminating Decimals

Both .5 and .125 are examples of **terminating decimals**. In each example we were actually able to finish each division with a zero remainder.

Sometimes a common fraction gives us a decimal fraction that never ends. For example:

$$\frac{1}{3} \text{ as a division gives .333…}$$

$$\frac{1}{7} \text{ as a division gives .14285714287}$$

Repeating Decimals

We call $.3\overline{3}$ and $.\overline{142857}$ **repeating decimals**. A line is placed over the digits of the decimal that repeat. The divisions for $\frac{1}{3}$ and $\frac{1}{7}$ are shown below.

134   Developmental Arithmetic–A Computational Review

```
     .3333              .142857
  3)1.0000           7)1.0000000
    9                  7
   ──                 ──
    10                 30
     9                 28
    ──                 ──
    10                 20
     9                 14
    ──                 ──
    10                 60
                       56
                      ──
                       40
                       35
                      ──
                       50
                       49
                      ──
                       10
```

Notice that in the division for $\frac{1}{3}$ each time we add a zero, we get the same remainder of 1. Hence, we keep getting 3s in our answer. Since the 3s go on forever, we are not able to finish the division. So, we either put three dots after the last 3 we had in our division or we place a bar over the last 3 we write.

Use 3 Dots Or A Bar To Show A Decimal Is Repeating

$$\frac{1}{3} = .333\ldots \text{ or } .33\overline{3}$$

In the division $\frac{1}{7}$ we can do the same thing but we need to list the repeating digits twice before we place the 3 dots.

$$\frac{1}{7} = .142857142857\ldots \text{ or } \overline{.142857}$$

One final word on changing a common fraction to a decimal fraction: We will always get a decimal fraction that either stops (terminates) or repeats. Because of this we say that a common fraction can always be written as a repeating or terminating decimal. It is important to know that there are decimals that do <u>not</u> terminate and do <u>not</u> repeat. The

A Common Fraction Always Gives A Repeating Or Terminating Decimal Fraction

Lesson Fourteen—Decimal Fractions

Rational Numbers

Irrational Numbers

Real Numbers

Place Value

square root of 2, written as $\sqrt{2}$, is an example of such a decimal. If you decide to study algebra, you will find that the common fractions and whole numbers are referred to as **rational numbers**. They are what we are referring to as decimal fractions. Non-repeating, non-terminating decimals are called **irrational numbers** because they cannot be written as common fractions. Together the rational and irrational numbers become the **real numbers** that are studied in algebra. Now, back to our work on decimal fractions.

We again need to talk about **place value** (you may want to go back and review Lesson 1 on place value). Since we have a decimal point, we need to find the value of each position to the right of the decimal point.

When we worked with whole numbers, we had the place value as shown below:

| _____ | _____ | _____ | _____ | _____ |
|10,000|1,000|100|10|1|

$\longleftarrow \times 10$

Remember as we went from right to left, we multiplied by a 10. If we go from left to right (⟶) we see that we divide by 10 each time to get the place value of the digit to the right. Of course we end up with the last digit to the right as a 1. We then place a decimal point after the 1.

Now we will add positions to the right of the decimal and **we will continue dividing by 10 each time as we go to the right (** ⟶ **).** We divide 1 by 10 and get $\frac{1}{10}$. So the first position to the right of the decimal point has a value of $\frac{1}{10}$.

Developmental Arithmetic–A Computational Review

We find the value of the second digit to the right by again dividing by 10.

$$\frac{1}{10} \div 10 = \frac{1}{10} \times \frac{1}{10} = \frac{1}{100}$$

The place value of the second digit to the right of the decimal point is $\frac{1}{100}$. We continue in this way to get the place value of each new position. The **place value chart** now looks like this:

Place Value Chart

etc. ← 1,000   100   10   1 $\frac{1}{10}$ $\frac{1}{100}$ $\frac{1}{1,000}$ → etc.

We can now talk about the **place value of a digit** in a decimal fraction. For example, in the decimal fraction .125, the digit

The Place Value Of A Digit

    1 is in the tenths position
    2 is in the hundredths position
    5 is in the thousandths position

As before, we can use expanded notation to write a decimal fraction. We write .125 as follows:

Writing A Decimal Fraction Using Expanded Notation

$$1 \times \frac{1}{10} + 2 \times \frac{1}{100} + 5 \times \frac{1}{1,000} \quad \text{or} \quad \frac{1}{10} + \frac{2}{100} + \frac{5}{1,000}$$

Lesson Fourteen–Decimal Fractions

Using Verbal Notation For A Decimal Fraction In Positional Notation

We also **use verbal notation to read a decimal aloud**. We use the place value of the "right most" digit. For example, the place value of the 5 in .125 is $\frac{1}{1,000}$ or thousandths. We read .125 as one hundred twenty-five thousand<u>ths</u>. Notice the "ths" at the end of the thousand.

The decimal fraction 2.125 is read by first saying the whole number and then the decimal fraction. We say "two and one hundred twenty five thousandths." Note that this is the **only** time we use the word "and" when verbalizing numbers: at the decimal point position.

Sometimes we need **to write the positional notation for a decimal fraction given in verbal notation**. For example, let's write the following number in positional notation: two and twenty-five thousandths. We will need thousandths, so we draw blanks for each position up to and including thousandths.

Using Positional Notation For A Decimal Fraction In Verbal Notation

$$\underline{} \, . \, \underline{} \, \underline{} \, \underline{}$$
$$1 \qquad \frac{1}{10} \quad \frac{1}{100} \quad \frac{1}{1,000}$$

Now we fill in the blanks, making sure our last blank has the last digits of our number. Notice that we have to fill in a space with a zero as a place holder.

$$\underline{2} \, . \, \underline{0} \, \underline{2} \, \underline{5}$$
$$1 \qquad \frac{1}{10} \quad \frac{1}{100} \quad \frac{1}{1,000}$$

REVIEW

- **Changing From a Common Fraction to a Decimal Fraction** *See page 133*
  We change a common fraction to a decimal fraction by dividing the denominator into the numerator. A decimal point must be placed in the dividend before adding zeros. A decimal point is placed in the answer and the division is carried out until the decimal either terminates or starts repeating.

- **Form of Common Fractions When Written as a Decimal Fraction** *See page 135*
  A decimal fraction either terminates, such as .25 or repeats, such as .3$\overline{3}$. We call such decimals rational numbers. Decimals that do not repeat or terminate, such as $\sqrt{2}$, are called irrational numbers. Together the rational and irrational numbers belong to what is called the set of real numbers.

- **Place Value With Decimal Fractions** *See page 137*
  We get the place value of the digits to the right of the decimal point by dividing by 10 each time. We must be careful to add a "ths" at the end of each place value; for example we must write hundredths rather than hundred for the place value of the 5 in the decimal fraction .25.

$$\underline{}\,.\,\underline{}\,\underline{}\,\underline{}$$

ones · tenths · hundredths · thousandths

Lesson Fourteen–Decimal Fractions

EXERCISES

1. Write each of the following common fractions as decimal fractions. (Do the division for each one.)

   a. $\frac{1}{4}$ b. $\frac{3}{5}$ c. $\frac{2}{3}$

   d. $\frac{3}{8}$ e. $\frac{1}{16}$ f. $\frac{2}{7}$

2. Write each of the following using expanded notation.

   a. .75   b. .136

   c. .005   d. 3.25

   e. 1.02   f. 51.1275

3. In the decimal fraction 13.0675, name the digit in the:
   a. Hundredths position
   b. Ten thousandths position
   c. Tenths position
   d. Ones position
   e. Thousandths position
   f. Tens position

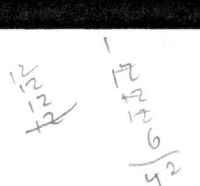

4. Write the following in verbal notation (remember to use "ths").

   a. .6

   b. .275

   c. 3.162

   d. 127.25

   e. .00125

   f. 5.075

5. Write each of the following using positional notation:

   a. two and five tenths

   b. one hundred twenty-five ten thousandths

   c. twenty-five and three hundred seventy-five thousandths

   d. one and thirty-five millionths

   e. one thousand three and two hundredths

   f. two hundred three ten thousandths

Lesson Fourteen–Decimal Fractions

ANSWERS TO EXERCISES

1. a. .25   b. .6   c. .6$\overline{6}$

   d. .375   e. .0625   f. .$\overline{285714}$

2. a. $7 \times \frac{1}{10} + 5 \times \frac{1}{100}$

   b. $1 \times \frac{1}{10} + 3 \times \frac{1}{100} + 6 \times \frac{1}{1,000}$

   c. $0 \times \frac{1}{10} + 0 \times \frac{1}{100} + 5 \times \frac{1}{1,000}$

   d. $3 \times 1 + 2 \times \frac{1}{10} + 5 \times \frac{1}{100}$

   e. $1 \times 1 + 0 \times \frac{1}{10} + 2 \times \frac{1}{100}$

   f. $5 \times 10 + 1 \times 1 + 1 \times \frac{1}{10} + 2 \times \frac{1}{100} + 7 \times \frac{1}{1,000} + 5 \times \frac{1}{10,000}$

3. a. 6   b. 5   c. 0

   d. 3   e. 7   f. 1

4. a. six tenths

   b. two hundred seventy-five thousandths

   c. three and one hundred sixty-two thousandths

   d. one hundred twenty-seven and twenty-five hundredths

   e. one hundred twenty-five hundred thousandths

   f. five and seventy-five thousandths

5. a. 2.5   b. .0125

   c. 25.375   d. 1.000035

   e. 1,003.02   f. .0203

Lesson Fourteen—Decimal Fractions

Quiz 2 Review

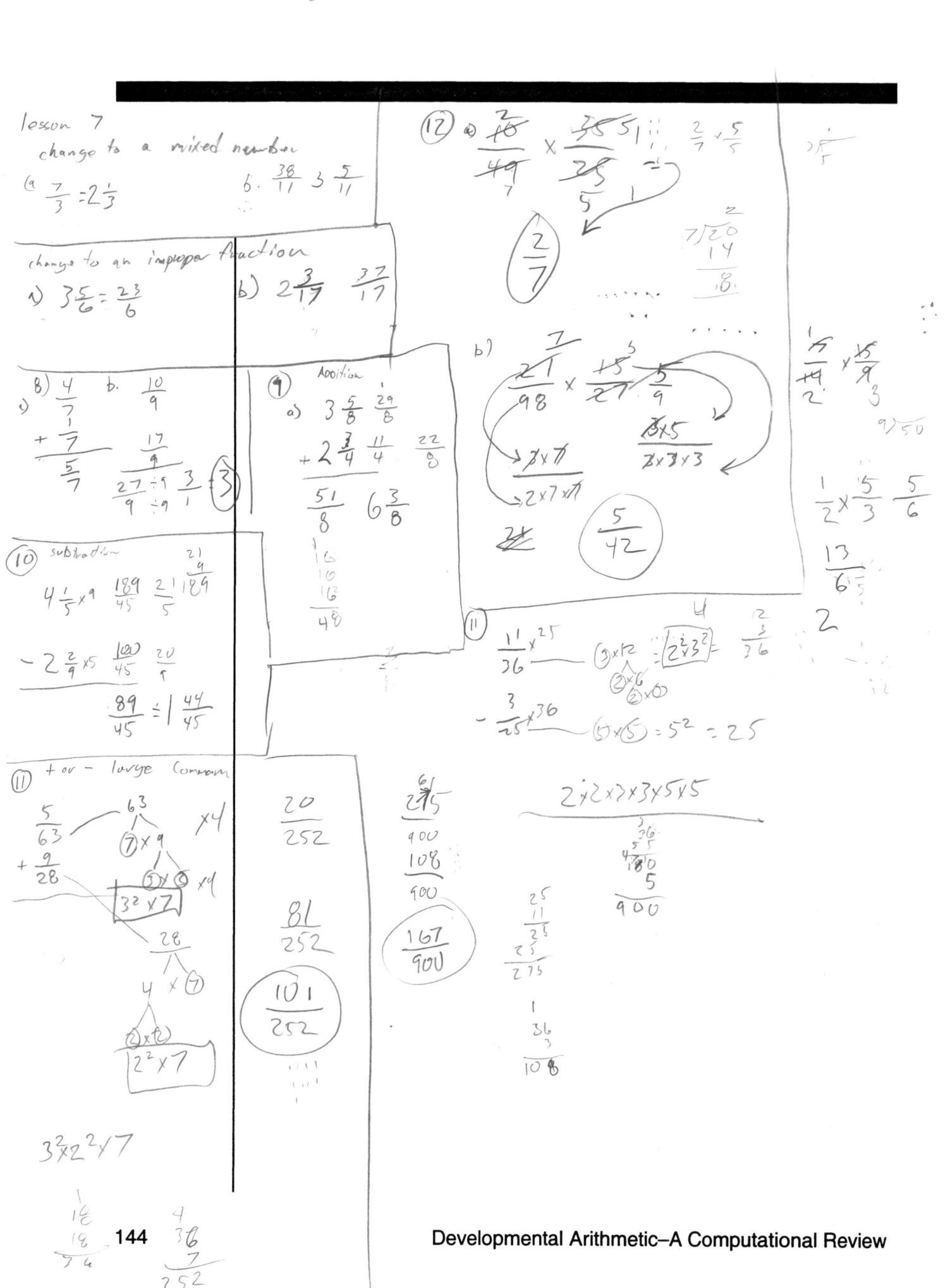

Lesson 7
change to a mixed number
a) $\frac{7}{3} = 2\frac{1}{3}$ b) $\frac{38}{11} = 3\frac{5}{11}$

change to an improper fraction
a) $3\frac{5}{6} = \frac{23}{6}$ b) $2\frac{3}{17} = \frac{37}{17}$

8) a) $\frac{4}{7} + \frac{1}{7} = \frac{5}{7}$ b) $\frac{10}{9} + \frac{17}{9} = \frac{27}{9} \div \frac{9}{9} = \frac{3}{1} = 3$

9) Addition
a) $3\frac{5}{8} = \frac{29}{8}$
 $+ 2\frac{3}{4} = \frac{11}{4} = \frac{22}{8}$
 $= \frac{51}{8} = 6\frac{3}{8}$

10) Subtraction
$4\frac{1}{5} \times 9 = \frac{189}{45} = \frac{21}{5} = \frac{21\frac{9}{9}}{189}$
$- 2\frac{2}{9} \times 5 = \frac{100}{45} = \frac{20}{9}$
$= \frac{89}{45} = 1\frac{44}{45}$

11) for — large common
$\frac{5}{63}$
$+ \frac{9}{28}$

$63 = 7 \times 9 = 3^2 \times 7$
$28 = 4 \times 7 = 2^2 \times 7$

$3^2 \times 2^2 \times 7$

12) a) $\frac{2}{\cancel{10}} \times \frac{\cancel{35}}{\cancel{49}} = \frac{2}{7} \times \frac{5}{5} = \boxed{\frac{2}{7}}$

b) $\frac{7}{\cancel{21}} \times \frac{\cancel{15}}{\cancel{27}} = \frac{5}{42}$
$\frac{3 \times 5}{2 \times 3 \times 3}$
$\boxed{\frac{5}{42}}$

$\frac{\cancel{5}}{\cancel{14}} \times \frac{\cancel{15}}{\cancel{9}} = \frac{1}{2} \times \frac{5}{3} = \frac{5}{6}$

$\frac{11}{36} \times 25$
$- \frac{3}{25} \times 36$

$\frac{20}{252}$
$\frac{81}{252}$
$\boxed{\frac{101}{252}}$

$\frac{6\frac{1}{3}}{275}$
$\frac{108}{900}$
$\boxed{\frac{167}{900}}$

$2 \times 2 \times 2 \times 3 \times 5 \times 5$

144 Developmental Arithmetic–A Computational Review

lesson fifteen

ADDITION AND SUBTRACTION OF DECIMAL FRACTIONS

The addition of decimal fractions is rather easy. You should remember that when we added whole numbers, we had to be careful to line up the digits so that they were in the correct place value columns. We did this by lining up the digits in the one's place.

When we add decimal fractions, we must again be careful to add like place values. We do this with decimal fractions by **lining up the decimal points**. We then guarantee that tenths are added to tenths, hundredths are added to hundredths, and so on. For example, 3.12 + .046 is set as follows:

Line Up The Decimal Points To Add Decimal Fractions

$$\begin{array}{r} 3.12 \\ + .046 \end{array}$$

It is sometimes easier if we first make sure each decimal fraction has the same number of digits after the decimal point. In the above example we can rewrite 3.12 as 3.120 because adding 0 thousandths does not change the value of 3.12. We then add as shown below. We must be careful to place a decimal point in the answer so it lines up with the decimal points in the problem.

$$\begin{array}{r} 3.120 \\ + .046 \\ \hline 3.166 \end{array}$$

Lesson Fifteen–Addition and Subtraction of Decimal Fractions

The placement of additional zeros makes some problems easier to handle. As an example we will look at the following addition problem.

$$1.235 + .41 + 23.2 + 326.07$$

We first change each decimal by adding zeros so that there are three digits after the decimal point.

$$1.235 + .410 + 23.200 + 326.070$$

Now we line up the decimal points and add. Again, notice that the decimal point in the answer is lined up with the decimal points in the problem. If the decimal point is left out of the answer, the error is a very serious error; so be careful to place the decimal point in the correct place in the answer. **Just remember, all of the decimal points must line up.**

Adding Decimal Fractions

```
   1.235
    .410
  23.200
+326.070
─────────
 350.915
```

Subtracting Decimal Fractions

We subtract decimal fractions as easily as we add them. We again **line up the decimal points and subtract**; being careful to line up the decimal point in the answer with the decimal points in the problem. The following subtraction illustrates what should be done.

$$15.06 - 12.478$$

We <u>must</u> add zeros to get the same number of digits after the decimal point for each decimal fraction. In subtraction it is essential that we add the zeros in case we have to borrow. We then line up the decimal points and subtract.

```
       4  9 5
    1 5. 0 6 0
  - 1 2. 4 7 8
  ─────────────
      2. 5 8 2
```

Whether we are doing additions or subtractions, once the decimal points are taken care of we can add and subtract as though the decimal point was not there. Of course, we must be careful to place the decimal point in the answer to line up with the decimal points in the problem.

We are often asked to add whole numbers to a decimal fraction. For example, we might be asked to find the sum 15 dollars and $1.75.

$$15 + 1.75$$

Since every whole number can be written with a decimal, we write 15 as 15.00. We can now line up the decimal points and add as follows:

Be Careful To Line Up The Deciaml Point In The Answer With The Decimal Point In The Problem

$$\begin{array}{r} 15.00 \\ +\ 1.75 \\ \hline 16.75 \end{array}$$

If the above problem were a subtraction problem, it would become even more critical that the additional zeros be written in. If the problem were left in the form

$$\begin{array}{r} 15. \\ -\ 1.75 \end{array}$$

When Subtracting Decimal Fraction, Additional Zeros Should Be Written In

there is the danger of forgetting to borrow and incorrectly bringing down the .75 in the answer. With the zeros written in it is clear that one needs to borrow; getting the correct answer that involves a .25, as is shown below.

$$\begin{array}{r} 1\overset{4}{\cancel{5}}.\overset{9}{\cancel{0}}{}^{1}0 \\ -\ 1.75 \\ \hline 13.25 \end{array}$$

Thus far it appears that we can add zeros at the end of a decimal fraction without any problems. This is not completely true. The number of digits after the decimal point provides us information about how precise a measurement is when the decimal fraction is the result of measuring something. In such instances, adding zeros may give a false sense of the precision used. The measurement .5 tells us that the

Lesson Fifteen–Addition and Subtraction of Decimal Fractions

Rounding Off

measurement was done precisely to the nearest tenth. If the measurement is written as .50, we conclude that the measurement was made precisely to the nearest hundredth. As a result, we need to be careful to ask if the decimal fractions are from measurements someone made. If they are, we must be careful to do what we call **rounding off** to the least precise measurement. For example, if we are asked to add 15 + 1.75 as measurements, we should give the answer to the nearest unit because the 15 is the least precise measurement. Note that the 15 has the precision of being written to the nearest one; whereas, the 1.75 is precise to the nearest hundredth. The answer we got earlier needs to be rounded off to the nearest unit; 16.75 is closer to 17 than it is to 16, so we say that 16.75 is rounded up to 17, the nearest unit or one's place.

Anyone who does his or her own income tax knows that the instructions say that one should round off to the nearest dollar. So, if one has a deduction of $327.16, the value should be rounded off to $327. If however, one has a deduction of $327.85, the value is rounded off to $328. The question, of course, is how does one decide whether to round up or round down.

Before we look at the rules for rounding off, it is important to know that we can round off both whole numbers and decimal fractions. The rules for each are just a little different. However, in both situations it is necessary to know what precision is required in the answer. Usually someone indicates the level of precision required, as in our example about taxes. At other times we may have to make the decision ourselves.

Rule Of Thumb For Rounding Off

| **As a rule of thumb, round off to the nearest hundredth.** |

The rule of thumb provides answers that work well with most problems whether they involve money or the distance between two points. If in doubt about the level of precision required, ask whoever gave you the problem.

The following rule works for rounding off numbers to the one's place and any place to the right of the decimal point.

Rules For Rounding Off To The Right Of The Decimal Point

First decide the place you want to round off to. Then
1. Look at the digit to the right of the place you want to round off to.
2. If that digit is a 1,2,3,or 4, you only need to write the digits to the left of that digit as they are. The other digits are dropped.
3. If that digit is a 5,6,7,8,or 9, the digit in the position you are rounding off to is increased by 1. The other digits are dropped.

Example: Round off 14.7136 to the

one's place

    1. The digit to the right of 4 is 7.
    2. Round off to 15.

tenth's place

    1. The digit to the right of 7 is 1.
    2. Round off to 14.7

hundredth's place

    1. The digit to the right of 1 is 3.
    2. Round off to 14.71

thousandth's place

    1. The digit to the right of 3 is 6.
    2. Round off to 14.714

When we round off numbers to the right of the decimal point we simply drop the digits we do not want. When we round off numbers to the left of the decimal point, we often have to replace digits with zeros.

Lesson Fifteen–Addition and Subtraction of Decimal Fractions

The following rule works for rounding off numbers to the left of the decimal point.

First decide the place you want to round off to. Then
1. Look at the digit to the right of the place you want to round off to.
2. If that digit is a 1, 2, 3, or 4, you only need to write the digits to the left of that digit as they are; replace the digits between that digit and the decimal point with zeros. Any digits after the decimal point are dropped.
3. If that digit is a 5, 6, 7, 8, or 9, the digit in the position you are rounding off to is increased by 1; replace the digits between that digit and the decimal point with zeros. Any digits after the decimal point are dropped.

Rules For Rounding Off To The Left Of The Decimal Point

Example: Round off 43,526.13 to the
ten's place
    1. The digit to the right of 2 is 6.
    2. Round off to 43,530

hundred's place
    1. The digit to the right of 5 is 2.
    2. Round off to 43,500

thousand's place
    1. The digit to the right of 3 is 5.
    2. Round off to 44,000

ten thousand's place
    1. The digit to the right of 4 is 3.
    2. Round off to 40,000

REVIEW

- **Adding Decimal Fractions** *See page 145*
  Line up the decimal points and add, being careful to line up the decimal point in the answer with the decimal points in the problem.

- **Subtracting Decimal Fractions** *See page 146*
  Line up the decimal points and subtract, being careful to line up the decimal point in the answer with the decimal points in the problem.

- **Rounding Off** *See page 149*
  The following rule works for rounding off to the one's place and any place to the right of the decimal point.
  First decide the place you want to round off to and
  1. Look at the digit to the right of the place you want to round off.
  2. If that digit is a 1, 2, 3, or 4, you only need to write the digits to the left of that digit as they are. The other digits are dropped.
  3. If that digit is a 5, 6, 7, 8, or 9, the digit in the position you are rounding off to is increased by 1. The other digits are dropped.

  The following rule works for rounding off to the left of the decimal point.
  First decide the place you want to round off to and
  1. Look at the digit to the right of the place you want to round off.
  2. If that digit is a 1, 2, 3, or 4, you only need to write the digits to the left of that digit as they are; replace the digits between that digit and the decimal point with zeros. Any digits after the decimal point are dropped.
  3. If that digit is a 5, 6, 7, 8, or 9, the digit in the position you are rounding off to is increased by 1; replace the digits between that digit and the decimal point with zeros. Any digits after the decimal point are dropped.

EXERCISES

1. Add the following decimal fractions.

   a. 1.706 + 3.467
   $$\begin{array}{r} \overset{1\ \ 1}{1.706} \\ 3.467 \\ \hline 5.173 \end{array}$$

   b. 27.049 + 8.12
   $$\begin{array}{r} \overset{1}{27.049} \\ 8.120 \\ \hline 35.169 \end{array}$$

   c. .00686 + 3.417
   $$\begin{array}{r} .00686 \\ 3.41700 \\ \hline 3.42386 \end{array}$$

   d. 4 + 1.36
   $$\begin{array}{r} 4. \\ 1.36 \\ \hline 5.36 \end{array}$$

2. Add the following decimal fractions.

   a. 1.735 + .38 + 14.12 + .369
   $$\begin{array}{r} .380 \\ 14.120 \\ .369 \end{array}$$

   b. .006 + 3.1 + 8.76 + 15.038

   c. 768.34 + 1.609 + 77.5 + 19.703

   d. 4,603.27 + 163.8 + 2,100.075

3. Subtract the following decimal fractions.

   a. 4.107 − 2.34

   b. .378 − .069

   c. 14.06 − 12.789

   d. 20.303 − .6547

4. Round off each number as indicated.

   a. 1.276 to the nearest tenth.

   b. 14.78 to the nearest unit or one's position.

   c. .0765 to nearest thousandth.

   d. 147.4 to the nearest ten.

   e. 3.14159 to the nearest hundredth.

   f. 1.4142135 to the nearest hundred thousandth.

   g. 1,379.58 to the nearest hundred.

   h. .142857 to the nearest ten thousandth.

ANSWERS TO EXERCISES

1.
    a. 5.173   b. 35.169   c. 3.42386   d. 5.36

2.
    a. 16.604   b. 26.904   c. 867.152   d. 6,867.145

3.
    a. 1.767   b. .309   c. 1.271   d. 19.6483

4.
    a. 1.3   b. 15   c. .077   d. 150

    e. 3.14   f. 1.41421   g. 1,400   h. .1429

lesson sixteen

MULTIPLICATION OF DECIMAL FRACTIONS

When multiplying decimal fractions, we follow a simple rule for placing the decimal point. The following discussion illustrates the rule.

Let's multiply $.23 \times 5.9$. First let's look at the multiplication as though we were multiplying two common fractions.

- We write each decimal as a common fraction:

$$.23 = \frac{23}{100} \qquad 5.9 = 5\frac{9}{10} = \frac{59}{10}$$

- We multiply the fractions:

$$\frac{23}{100} \times \frac{59}{10} = \frac{1,357}{1,000}$$

- We change $\frac{1,357}{1,000}$ to the mixed number $1\frac{357}{1,000}$.

- We say the number aloud,

  "one and three hundred fifty-seven thousandths."

- Finally, we use positional notation to write what we heard:

  1.357

Notice that there are three digits to the right of the decimal point. There are two digits to the right of the decimal in .23 and one digit to the right of the decimal in 5.9. This information suggests a rule to locate the decimal point when two numbers with decimals are multiplied.

Rule For Multiplying Decimals

Locating the decimal point when two decimals are multiplied:

1. Multiply the numbers as though there were no decimal points.
2. Count the number of digits to the right of the decimal point in each factor (each number).
3. Add the two counts in step 2.
4. Position the decimal point so that there are that many digits to the right of the decimal point in the final answer. Count from the right to the left (◄———) because we are dividing by a power of ten.

Now let's use the rule to find the product of .23 × 5.9

1.
```
        2
       5.9
    ×  .23
      ─────
       177
      118
      ─────
      1357
```

2. There are two digits to the right of the decimal in .23
   There is one digit to the right of the decimal in 5.9

3. 2 + 1 = 3 So we will be dividing by 10^3.

4. Count 3 digits from the right and place the decimal

   here
   ↓
   1.357

As a second example, let's multiply 2.023 × 14.783

1.
```
            6 2
         14.783
      ×   2.023
         ──────
          44349
         29566
        00000
       29566
       ────────
       29906009
```

2. There are three digits to the right of the decimal in 14.783
   There are three digits to the right of the decimal in 2.023

156 Developmental Arithmetic–A Computational Review

3. $3 + 3 = 6$ We will be dividing by 10^6 and move the point 6 places.

4. Count 6 digits from the right and place the decimal
here
↓
29.906009

Let's look at another example:

$$45.2 \times .0003$$

Remember that multiplication can be done in any order. We will multiply this problem both ways.

Multiplication Can Be Done In Any Order

1.   .0003 (4 digits)
    × 45.2 (1 digit)
        0006
       0015
      0 012
      0.01356 (need 5 digits to the right of the decimal)

2.   45.2 (1 digit)
    × .0003 (4 digits)
      .01356

Obviously, the second way of doing the problem was shorter as it was not necessary to multiply by the zeros. We did, however, need to put a 0 in the answer as a place holder to get five digits after the decimal point. Note that the 0 went before the other digits, <u>not</u> at the end.

We can multiply whole numbers by a decimal fraction. Remember, every whole number can be written as a decimal. 7×1.25 can be written as $7. \times 1.25$. Let's look at the multiplication done both ways.

Multiplying Whole Numbers By A Decimal Fraction

1. 1.25 (2 digits)   2. 1.25 (2 digits)
   × 7.   (0 digits)   × 7   (0 digits)
    8.75   8.75

Since $2 + 0 = 2$, we need 2 digits to the right of the decimal point in either answer. As you can see, it is easier not to write the decimal after the whole number. The result is the same.

Lesson Sixteen—Multiplication of Decimal Fractions

A Short Cut For Multiplying By A Power Of Ten

You may remember that there is an easy way of multiplying a whole number by a power of ten like 10 or 100 or 1000, etc. We used the short cut in Lesson 4 when we multiplied whole numbers by a power of ten. We just add the correct number of zeros to the end of the number we are multiplying. For example: $100 \times 24 = 2400$; all we did was write two zeros after the 24.

Now that we are working with decimals, we can look at the problem in a different way. The number 24 can be written as 24.00 by placing the decimal point and adding zeros. Likewise, 2400. = 2400. When we multiply 24.00 by 100 we simply move the decimal point two places to the right to get 2400. If we write the product as 24×10^2, it is even easier to use the shortcut. The exponent of 2 tells us to move the decimal point 2 places to the right.

The previous example illustrates two rules for multiplying a decimal fraction by a power of ten:

Rules For Multiplying By A Power Of Ten

1. When a decimal fraction is multiplied by a 1 followed by zeros, move the decimal point one place to the right for each zero.

2. When a decimal fraction is multiplied by a power of ten, move the decimal point to the right the number of places given by the exponent.

It may be necessary to add additional zeros in some problems as the following example shows:

$$1{,}000 \times 4.17 \quad \text{or} \quad 10^3 \times 4.17$$

We need to move the decimal three places to the right. We rewrite 4.17 as 4.170 and then move the decimal. The answer is 4,170. Again, we needed to use a zero as a place holder.

REVIEW

- **Multiplication of Decimal Fractions** *See page 156*
  To multiply decimal fractions together we count the number of digits to the right of the decimal point in each factor. Add the two counts and position the decimal point in the answer so that there are as many digits to the right of the decimal point.

- **Multiplying By a Power of Ten** *See page 158*
  The shortcut for multiplying a decimal fraction by 1 followed by zeros is to move the decimal point in the decimal fraction one place to the right for each zero. If the power of ten is written with an exponent, then move the decimal point to the right the number of places indicated by the exponent.

EXERCISES

1. Tell how many digits should be after the decimal point in each answer; then, do each multiplication.

   a. .0135 × 4.17

   $$\begin{array}{r} 4.1700 \\ \times\ 0.0135 \\ \hline 20.8500 \\ 125.100 \\ 417.00 \\ \hline 562.9500 \end{array}$$

   b. 1.34 × 2.03

   $$\begin{array}{r} 1.34 \\ 2.03 \\ \hline 402 \\ 000 \\ 268 \\ \hline 2.6802 \end{array}$$

   c. 23.79 × .008

   $$\begin{array}{r} 23.79 \\ .008 \\ \hline .19032 \end{array}$$ (5)

   d. 402.68 × 17.4

   $$\begin{array}{r} 402.68 \\ 17.4 \\ \hline 161072 \\ 161072 \\ 10268 \\ \hline 2798.592 \end{array}$$ (3)

   e. 5.076 × .074

   $$\begin{array}{r} 5.076 \\ .074 \\ \hline 20304 \\ 35532 \\ 0000 \\ \hline .375624 \end{array}$$ (6)

   f. 4 × 3.75

   $$\begin{array}{r} 3.75 \\ 4 \\ \hline 15.00 \end{array}$$ (2)

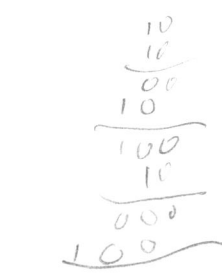

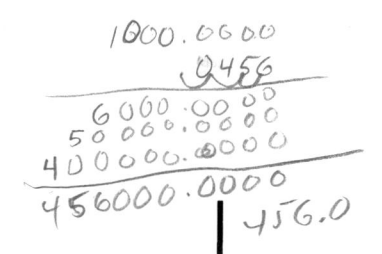

2. Find each product.

   a. 10×14.78

   b. $10^3 \times .0456$

   c. 100×14.6

   d. $10,000 \times 7.00654$

   e. $10^2 \times 23$

   f. $10^5 \times .4765$

Lesson Sixteen–Multiplication of Decimal Fractions

ANSWERS TO EXERCISES

1. a. 6 digits   .056295   b. 4 digits   2.7202

   c. 5 digits   .19032   d. 3 digits   7006.632

   e. 6 digits   .375624   f. 2 digits   15.00

2. a. 147.8   b. 45.6

   c. 1,460   d. 70,065.4

   e. 2,300   f. 47,650

lesson seventeen
DIVISION OF DECIMAL FRACTIONS

There are several ways of handling division problems that involve decimal fractions. Certainly the simplest is to use an inexpensive hand-held calculator, but there may at times be a need to do such problems long hand.

We have already reviewed the division process in two earlier lessons. In each lesson we discussed the need to work with decimals. When we reviewed division with whole numbers in Lesson 5, we said we would later use decimals to work with the remainders we often had left over in a division problem. In Lesson 14 where we reviewed changing a common fraction to a decimal fraction, we placed a decimal point at the end of the **dividend** (the number under the division sign), located the decimal point in the answer, and added zeros for each new digit we wanted after the decimal point in the answer. The following example illustrates the process. We will change $\frac{5}{4}$ to a decimal fraction with three digits after the decimal point.

Dividend

1. Place a decimal point after the 5. $4\overline{)5.}$
2. Add three zeros after the decimal point. $4\overline{)5.000}$
3. Place the decimal point in the answer. $4\overline{)5.000}$
4. Divide.
$$\begin{array}{r} 1.250 \\ 4\overline{)5.000} \\ \underline{4} \\ 1\,0 \\ \underline{8} \\ 20 \\ \underline{20} \\ 00 \end{array}$$

Changing A Common Fraction To A Decimal Fraction

You should first notice that the decimal point was placed in the answer directly above the point after the 5. We also had one more zero than we needed because the division stopped after the second digit. We can drop the last zero if we want to. If the partial remainders had continued, we could have added more zeros to 5.000 to carry out the answer further.

Often, the division does not stop. There may be a series of numbers that will repeat, in order, or perhaps it seems that the division will just continue. **If the answer begins to repeat digits, we can use either the three dots after one repeating digit or a bar on top of a series of digits that repeat.** Another solution is to stop the division, when you are satisfied that you have carried it far enough, and round off. You may want to review the section on rounding off in Lesson 15.

What To Do If The Division Does Not Stop

In all of the division that has been done with decimal fractions to this point, all we need to remember is to place the decimal in the answer directly above the decimal in the dividend. Now it is necessary to discuss when the divisor is a decimal fraction.

$$\text{divisor} \overline{\smash{)}\text{dividend}}^{\text{quotient (answer)}}$$

Rule for dividing when the divisor is a decimal fraction:

1. Move the decimal in the divisor until it is to the far right.
2. Count the number of places you moved the decimal.
3. Now move the decimal the same number of places to the right in the dividend. Place a decimal in the answer position directly above the decimal in the dividend.
4. Divide and add enough zeros so you have one more digit in the answer than you need.
5. If necessary, round off the answer.

Dividing By A Decimal Fraction

We will use the rule to do the following division:

$$.33\overline{).01386}$$

1. $.33.\overline{).01386}$

2. The decimal point was move two places to the right.

3. $33.\overline{).01.386}$

4.
$$33.\overline{)01.386} \;\; \begin{matrix} 00.042 \\ 1\,32 \\ \overline{66} \\ 66 \\ \overline{0} \end{matrix}$$

5. If we round off to the nearest hundredth, we have 0.04 as the answer. One zero is often written before the decimal point to remind the reader that we have a decimal.

The following explanation shows why the rule works. The division $.33\overline{).01386}$ is the same as the fraction $\dfrac{.01386}{.33}$. We can use the basic principle of fractions to multiply both the bottom and top of the fraction by 100 as follows:

Why The Division Rule Works

$$\frac{.01386}{.33} \times \frac{100}{100} = \frac{1.386}{33}$$

We can now write the fraction as the division $33\overline{)1.386}$ which is exactly what we obtained using the rule.

Lesson Seventeen—Division of Decimal Fractions

Let's go through another example. We will do the division

$$1.23\overline{)3.21}$$

1. $1.23.\overline{)3.21}$

2. The decimal was moved two places to the right.

3. $123\overline{)3.21.}$

4. $\begin{array}{r}2.\\123\overline{)321.}\\\underline{246}\\75\end{array}$

We need to add zeros to continue the division.

$$\begin{array}{r}2.609\\123\overline{)321.000}\\\underline{246}\\75\,0\\\underline{73\,8}\\1\,200\\\underline{1\,107}\\93\end{array}$$

5. We could add more zeros, but we will stop and round off the answer to 2.61 (to the nearest hundredth).

Sometimes division problems work out easily. The division $.006\overline{)18}$ is easy to do if we remember that 18 and 18. are the same. (A decimal can always be placed after a whole number.)

Let's do the division $.006\overline{)18.}$

Move the decimal point in the divisor and then in the dividend according to the rule.

$$.006.\overline{)18.000.}$$

Now do the division as follows:

$$\begin{array}{r} 3000. \\ 006.\overline{)18000.} \\ \underline{18} \\ 0000 \end{array}$$

Let's do an example using a whole number divisor so that we can learn a shortcut for some division.

$$\begin{array}{r} .0452 \\ 100\overline{)4.5200} \\ \underline{4\ 00} \\ 520 \\ \underline{500} \\ 200 \\ \underline{200} \\ 0 \end{array}$$

We divided 4.52 by 100 and we have .0452 for our answer. The answer has the same non-zero digits as 4.52 (the dividend). The decimal has been moved to the left two digits. This example **illustrates a shortcut to use when dividing by a power of ten** like 10 or 100 or 1,000.

Move the decimal in the dividend one digit to the left for each zero that is in the divisor. (It helps to remember that the answer will be smaller than the dividend.) **If the power of ten is written using an exponent, move the decimal point to the left the number of places given by the exponent.**

Dividing By Powers Of Ten

The following are some examples for you to study to become more familiar with the shortcut. Note that it is sometimes necessary to place zeros between the original digits and the decimal point.

$$100\overline{)5} = 0.05 \qquad \frac{413.76}{10} = 41.376$$

$$\frac{2.17}{100} = 0.0217 \qquad 1000\overline{)56.78} = 0.05678$$

$$10\overline{)40.13} = 4.013 \qquad \frac{.03}{100} = 0.0003$$

Lesson Seventeen—Division of Decimal Fractions

REVIEW

See page 163

- **Dividing By a Whole Number**
  If there is no decimal in the divisor, place the decimal point directly above the decimal in the dividend and proceed with division.

See page 164

- **Dividing By a Decimal Fraction**
  When there is a decimal in the divisor, move the decimal point to the far right. Count the number of places the point was moved and move the decimal point in the dividend the same number of places. Place a decimal in the answer position directly above the decimal in the dividend and proceed with division.

See page 167

- **Dividing By a Power of Ten**
  A shortcut for division with divisors that are powers of ten such as 10, 100, 1,000 is to move the decimal point as many places to the left as there are zeros in the divisor. If the power of ten is written using exponents, move the decimal to the left the number of places indicated by the exponent.

EXERCISES

1. Do each division problem (round to nearest hundredth).

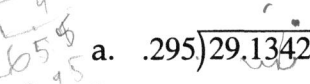

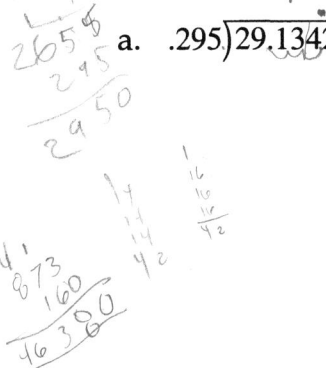

   a. .295)29.1342

   b. 7.56)142.8

   c. 87.3)110,434.5

   d. .065)31.512

   e. 1.27)4.3561

   f. .0003)1.2

Lesson Seventeen–Division of Decimal Fractions

2. Use the shortcut to do each division.

a. $31.46 \div 10$

b. $\dfrac{476.873}{1{,}000}$

c. $100\overline{).046}$

d. $617 \div 10^5$

e. $\dfrac{5.67}{10^3}$

f. $10\overline{)87.9}$

ANSWERS TO EXERCISES

Note: Answers for number 1 are rounded off to the nearest hundredth.

1. a. 98.76   b. 18.89

   c. 1,265.00   d. 484.80

   e. 3.43   f. 4,000.00

2. a. 3.146   b. .476873

   c. 0.00046   d. .00617

   e. 0.00567   f. 8.79

Lesson Seventeen—Division of Decimal Fractions

lesson eighteen
PERCENT

We have already reviewed common fractions and decimal fractions. In this lesson we complete our review of computational arithmetic by looking at a special form of common fractions. When we write a common fraction with a denominator of 100, we give the fraction a special name; we call the fraction a **percent.**

Percent

The percent symbol, %, was derived as a short-hand for representing a fraction with a denominator of 100. For example 25% is just a shortcut for writing the common fraction 25 over 100. The following example shows how the percent symbol was derived:

$$\frac{25}{100} = 25\big/100 = 25\big/{}^0_0 = 25\% = 25\%$$

This means that we now have three different forms that we can use to write a fraction:

- Common fraction   CF
- Decimal fraction   DF
- Percent   %

Three Forms Of A Fraction

Often it is necessary to change from one form of a fraction to a different form. For example we can write the fraction in the above example in the three different forms as follows:

$\frac{25}{100}$	.25	25%
Common Fraction	**Decimal Fraction**	**Percent**

CF → DF
Changing a Common Fraction to a Decimal Fraction

Regardless of which form of a fraction we start with, we need to be able to write the fraction in either one of the other two forms. Before we look at how to change common fractions and decimal fractions to percents, we need to review again some of the ideas we have already studied in the previous lessons.

To change a common fraction to a decimal fraction we simply use division and divide the denominator of the fraction into the numerator. We place the decimal point and add zeros as needed. For example:

$$\frac{1}{4} = 4\overline{)1.00}^{.25} \qquad \frac{3}{8} = 8\overline{)3.000}^{.375}$$

DF → CF
Changing Decimal Fractions to Common Fractions

To change a decimal fraction to a common fraction, we need to remember the place value of the digits to the right of the decimal point. We then write what we would say in verbal notation as a common fraction. For example, if the decimal fraction has two digits to the right of the decimal point, like in .45, we know the decimal is written as hundredths. We would write the common fraction by writing the digits 45 over a denominator of 100. A fraction such as .625 has three digits to the right of the decimal point, and we read the decimal as thousandths. So we would write .625 as the common fraction 625 over 1,000. In each example we would want to reduce the common fractions as follows:

$$.45 = \frac{45}{100} = \frac{9}{20} \qquad .625 = \frac{625}{1000} = \frac{5}{8}$$

DF → CF
Decimal Fraction to a Common Fraction Using Powers of Ten

Another way to change from a decimal fraction to a common fraction is to write the decimal over 1. Then multiply both the numerator and denominator by the power of ten that moves the decimal point in the numerator to the far right. For example:

$$\frac{.45 \times 10^2}{1 \times 10^2} = \frac{45}{100} = \frac{9}{20} \qquad \frac{.625 \times 10^3}{1 \times 10^3} = \frac{625}{1000} = \frac{5}{8}$$

Of course, **if the decimal fraction is a repeating decimal, we must be more careful to write the decimal fraction with two digits and a remainder after the decimal point.** For example the decimal .333... can be written as .33⅓ and then changed to a common fraction using either of the above methods. The following solution is done using powers of ten.

$$\frac{.33\frac{1}{3} \times 10^2}{1 \times 10^2} = \frac{33\frac{1}{3}}{100} = \frac{\frac{100}{3}}{\frac{100}{1}} = \frac{100}{300} = \frac{1}{3}$$

If we use the verbal notation method, we start by writing .33⅓ as $\frac{33\frac{1}{3}}{100}$ and complete the solution in the same way we did above.

The easiest way to change any decimal fraction to a common fraction is to simply memorize the most used "common fraction" to "decimal fraction" equivalents. The following table of equivalents may help if you are willing to take the time to memorize them.

Decimal Fraction	Common Fraction
$.50 =$	$\frac{1}{2}$
$.25 =$	$\frac{1}{4}$
$.75 =$	$\frac{3}{4}$
$.33\frac{1}{3} =$	$\frac{1}{3}$
$.66\frac{2}{3} =$	$\frac{2}{3}$

Now, we are ready to write both common fractions and decimal fractions as percents. If we start with a decimal fraction, it is very easy to change to a percent. We only need to remember that we want to get a

DF → CF
Changing A Repeating Decimal To A Common Fraction

Memorize The Table

Table of Decimal Equivalents

Lesson Eighteen Percent 175

DF → %
Decimal Fraction to a Percent

common fraction with a denominator of 100. The following example shows what we do.

$$.875 = \frac{.875 \times 100}{1 \times 100} = \frac{87.5}{100} = 87.5\%$$

Although there were three digits after the decimal point, we only multiplied by 100 so that we would have a denominator of 100. In contrast we write .875 as a common fraction by multiplying both the numerator and denominator by 1,000 to get rid of the decimal point.

If we want, the above method can be simplified by noting that when we wrote the decimal fraction as a percent we simply moved the decimal point two places to the right. We state the rule as follows:

Rule for Changing from a Decimal Fraction to a Percent

> **To change a decimal fraction to a percent:**
>
> **Move the decimal point two places to the right and add a %sign at the end.**

When we start with a common fraction, we can always **rewrite the fraction as a decimal fraction and use the above rule to get a percent**. There are times, however, when it is much easier to just **rewrite the common fraction with a new denominator of 100**. We can then write the result with a percent sign as shown in the following example:

CF → %
Changing from a Common Fraction to a Percent

$$\frac{1}{4} = \frac{1 \times 25}{4 \times 25} = \frac{25}{100} = 25\%$$

This shortcut does not always work this well because we might have a fraction like 3 over 8. The shortcut only works when it is easy to change the denominator to 100. When we start with a fraction like $3/8$, it is best to **change the common fraction to a decimal fraction and move the decimal point two places to the right before adding a percent sign.**

Often times we need to change a percent back to a decimal fraction. The easiest way to do this is to reverse the procedure we use to get the percent. We just move the decimal point two places to the left and drop the % sign.

> **To change from a percent to a decimal fraction:**
>
> **Move the decimal point two places to the left and drop the percent sign.**

Rule for Changing from a Percent to a Decimal Fraction

The only difficulty with using the above rule for changing from a percent to a decimal fraction is remembering which way to move the decimal point. We also must remember that if we do not see a decimal point in the percent, the decimal point is understood to be at the far right of the number. The following example shows an alternate method that we can use to change from a percent to a decimal fraction.

% → DF

Changing from a Percent to a Decimal Fraction

$$85\% = \frac{85}{100} = \frac{85}{10^2} = .85$$

All we need to remember is that percent means over 100. So 85% is written as 85 over 100 or 10^2. We then need the rule that we used earlier for dividing by a power of ten.

> **When we divide by a power of ten, we move the decimal point to the left the number of places in the exponent.**

Rule for Dividing by a Power of 10

In the above example, we are dividing by 10^2. The exponent is a 2; so we move the decimal point two places to the left to get our decimal fraction.

There are a few percent problems involving fractions that can be difficult to change back into decimal fractions. We will look at two examples of percents that we come across when we work with banking and interest rates. We need to change each of the percents to a decimal fraction.

$\frac{1}{2}\%$ $\qquad$ $5\frac{1}{4}\%$

% → DF
Changing a Percent Involving Fractions to a Decimal Fraction

We need to be careful that we do not make the common error in changing $\frac{1}{2}\%$ to a decimal fraction. The error occurs when $\frac{1}{2}$ is first changed to the decimal .5 and the % is not written in. There is still a percent sign in the problem because all we have done at this point is change the fraction to a decimal fraction; after we change the fraction to the decimal fraction, we have .5% which we now rewrite as

$$\frac{.5}{100} = \frac{.5}{10^2} = .005$$

When we change $5\frac{1}{4}\%$ to a decimal fraction, we first write the $\frac{1}{4}$ as a decimal fraction to get 5.25%; which we then change to a decimal fraction by moving the decimal point two places to the left.

$$5.25\% = \frac{5.25}{100} = \frac{5.25}{10^2} = .0525$$

There are times that we may want to change from a percent to a common fraction. For example, if a store advertises a sale at 25% off the retail price, we might find it useful to know that this is the same as $\frac{1}{4}$ off the price. The simplest way to change from a percent to a common fraction is to write the percent as a fraction over 100. The following two examples show how we change a percent into a common fraction.

% → CF
Changing from a Percent to a Common Fraction

25% $\qquad$ $5\frac{1}{4}\%$

The first example is done easily by simply writing the number over 100; and then reducing the fraction to lowest terms.

$$25\% = \frac{25}{100} = \frac{1}{4}$$

In the second example we have two choices; we can either write the mixed number as a decimal fraction or we can write the mixed number as an improper fraction. The two solutions are shown below, but the improper fraction method has the advantage of giving a common fraction that is already reduced.

$$5\tfrac{1}{4}\% = \frac{5.25 \times 100}{100 \times 100} = \frac{525}{10,000} = \frac{21}{400} \quad \text{or} \quad 5\tfrac{1}{4}\% = \frac{\frac{21}{4}}{\frac{100}{1}} = \frac{21}{400}$$

% → CF
Changing from a Percent to a Common Fraction

REVIEW

See page 174
- **Changing a common fraction to a decimal fraction**
  Divide the denominator into the numerator; adding as many zeros as needed.

See page 174
- **Changing a decimal fraction to a common fraction**
  Remove the decimal point and put the digits over a denominator of 1 followed by a number of zeros equal to the number of digits to the right of the decimal point.

See page 176
- **Changing a decimal fraction to a percent**
  Locate the decimal point; move the decimal point two places to the right; and add a percent sign.

See page 176
- **Changing a common fraction to a percent**
  ❖ Change the common fraction to a new fraction with a denominator of 100. The numerator of the new fraction is then written; followed by the percent sign.
  or
  ❖ Change the common fraction to a decimal fraction and move the decimal point two places to the right; adding the percent sign at the end.

See page 177
- **Changing a percent to a decimal fraction**
  Remove the percent sign; then move the decimal point two places to the left.

See page 178
- **Changing a percent to a common fraction**
  Write the number without the percent sign over a denominator of 100; then simplify the result so it is a common fraction in reduced form.

EXERCISES

1. Write the following common fractions as a decimal fractions.

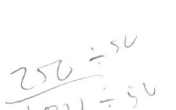

   a. $\frac{7}{8}$ .875   b. $\frac{2}{3}$ .666̄   c. $\frac{3}{4}$ .75   d. $\frac{5}{6}$ .833   e. $\frac{1}{5}$ .20

2. Write the following decimal fractions as common fractions.

   a. .25 $\frac{1}{4}$ b. .375 $\frac{37\frac{1}{2}}{100}$ c. .66$\frac{2}{3}$ d. .6 $\frac{60}{100}$ e. .005 $\frac{1}{2}$

3. Write the following decimal fractions as percents.

   a. .025 2.5%   b. .45 45%   c. .33$\frac{1}{3}$ 33.333%   d. 1.25 125%   e. .3 30%

4. Write the following percents as decimal fractions.

   a. 35% .35   b. $\frac{1}{4}$% .25   c. 150% 1.50   d. 2% .002   e. 6$\frac{1}{4}$% .0625

5. Write the following percents as common fractions

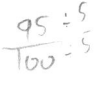

   a. 95% $\frac{19}{20}$ b. 250%   c. 33$\frac{1}{3}$%   d. $\frac{2}{3}$%   e. 7$\frac{1}{2}$%

6. Write the following common fractions as percents.

   a. $\frac{3}{8}$ b. $\frac{5}{2}$ c. $\frac{3}{5}$ d. $\frac{1}{100}$ 1%   e. $\frac{1}{30}$

Lesson Eighteen Percent

ANSWERS TO EXERCISES

1.
   a. .875   b. .66$\frac{2}{3}$ c. .75   d. .83$\frac{1}{3}$ e. .2

2.
   a. $\frac{1}{4}$ b. $\frac{3}{8}$ c. $\frac{2}{3}$ d. $\frac{3}{5}$ e. $\frac{1}{200}$

3.
   a. 2.5%   b. 45%   c. 33$\frac{1}{3}$%   d. 125%   e. 30%

4.
   a. .35   b. .0025   c. 1.5   d. .02   e. .0625

5.
   a. $\frac{19}{20}$ b. $\frac{5}{2}$ c. $\frac{1}{3}$ d. $\frac{1}{150}$ e. $\frac{3}{40}$

6.
   a. 37.5%   b. 250%   c. 60%   d. 1%   e. 3$\frac{1}{3}$%

REVIEW TEST 1

Name_____

1. Write the following numerals in words:
   a. 35,419

   b. 3,012,357

2. Write the following in expanded form using powers of ten:
   a. 7,138

   b. 59,203,075

3. Write the following numerals in positional notation:
   a. Five hundred thirty-two thousand three hundred seventeen

   b. Thirty- seven million eight hundred three thousand twelve

4. Given the numeral 9,258,307 write the place value of the:
   a. 3

   b. 0

5. Put commas in the following numerals:
   a. 138492

   b. 3060873562846

REVIEW TEST 2

Name_____

1. Find the following sums:

    a. 6 + 9   b. 5 + 0   c. 7 + 1   d. 9 + 8   e. 7 + 7

2. Use the low stress algorithm to find the following sums. Then check your answer for each column by reversing the addition.

    a. 2
        0
        6
        8
       +4
       ───

    b. 7
        5
        4
        9
       +3
       ───

3. Find the sums (first rewrite lining up the columns).
    a. 6,547 + 385 + 72,489 + 31 + 2,096

    b. 7,075 + 797 + 49,462 + 73 + 9

4. Use grouping by tens to find the sums for each problem.

    a. 74
        37
        86
        19
       +23

    b. 935
       362
       143
       476
      +724

5. Add, showing the carry for each column.

    a. 5,375
       1,983
      +7,258

    b. 7,659
         738
       8,942
      +   79

Review Tests

185

REVIEW TEST 3

Name_____

1. Find the following differences.

   a. 9 − 1   b. 8 − 5   c. 5 − 0   d. 7 − 7   e. 9 − 8

2. Subtract.

   a.   592
      −287

   b.   6,154
      −2,398

3. Subtract and show the check.

   a.   6,100,201
      −   765,832

   b.   40,805
      −12,938

4. Find the difference between each pair of numbers.

   a. 6,068 and 4,076   b. 90,006 and 63,529

5. Use the Austrian or Addition Method to find the differences.

   a.   5,007
      −2,489

   b.   9,752
      −4,598

REVIEW TEST 4

Name_____

1. a. 1 × 7 =   b. 7 × 6 =   c. 8 × 8 =   d. 0 × 5 =   e. 7 × 8 =

2. Multiply.

    a.   43
          ×21

    b.   57
          ×98

3. Find the products.

    a.   906
         × 85

    b.   1,257
         ×   860

4. Find the products.

    a.   6,873
        ×100,000

    b.   5,980
         ×1,000

5. Find the product of the three numbers.

    a. 5 × 4 × 7 =

    b. 0 × 8 × 9 =

REVIEW TEST 5

Name_____

1. Divide.

   a. $8\overline{)56}$ b. $\dfrac{0}{7}$ c. $81 \div 9$ d. $7\overline{)42}$ e. $12 \div 1$

2. Divide.

   a. $7\overline{)182{,}623}$ b. $76\overline{)1{,}528{,}056}$

3. Find the quotients.

   a. $1{,}134{,}972 \div 837$ b. $57\overline{)703{,}665}$

4. a. How many times can 235 be subtracted from 17,390?

   b. How many times can 144 be subtracted from 1,728?

5. Find the second of two factors.

   a. The first factor is 31 and the product is 7,967.

   b. The first factor is 8 and the product is 72.

Review Tests

REVIEW TEST 6

Name_____

Write each of the following as a product of primes using exponents.

1. 60

2. 75

3. 192

4. 171

5. 360

6. 468

7. 1,350

8. 2,340

9. 6,670

10. 12,435

REVIEW TEST 7

Name_____

1. Write as a product of primes using exponents.

    a. 450

    b. 2,310

2. Use prime factoring to reduce each fraction.

    a. $\dfrac{36}{54}$

    b. $\dfrac{48}{80}$

3. Rewrite the mixed number as an improper fraction.

    a. $2\frac{5}{7}$

    b. $3\frac{1}{4}$

4. Rewrite the improper fractions as a mixed number.

    a. $\dfrac{15}{4}$

    b. $\dfrac{23}{2}$

5. Do the division and write the remainder as a reduced fraction.

    a. $1{,}652 \div 12$

    b. $43{,}144 \div 45$

Review Tests

REVIEW TEST 8

Name_____

1. Use the Basic Principle of Fractions to change the fraction to a new one with the denominator specified.

   a. $\dfrac{3}{4} = \dfrac{}{12}$

   b. $\dfrac{5}{2} = \dfrac{}{16}$

2. Add.

   a. $\dfrac{3}{7} + \dfrac{2}{7}$

   b. $\dfrac{5}{9} + \dfrac{1}{9}$

3. Subtract.

   a. $\dfrac{4}{7} - \dfrac{3}{7}$

   b. $\dfrac{5}{8} - \dfrac{3}{8}$

4. Use prime factoring to reduce each fraction.

   a. $\dfrac{36}{48}$

   b. $\dfrac{120}{450}$

5. a. Write 1,260 as a product of primes.

   b. Find the product of $2 \times 2 \times 3 \times 5 \times 7$.

Review Tests

REVIEW TEST 9

Name _____

1. Add.

   a. $\dfrac{3}{7} + \dfrac{5}{7}$

   b. $\dfrac{1}{10} + \dfrac{3}{10}$

2. Subtract.

   a. $\dfrac{5}{9} - \dfrac{2}{9}$

   b. $\dfrac{5}{4} - \dfrac{3}{4}$

3. Add.

   a. $\dfrac{2}{5} + \dfrac{1}{3}$

   b. $\dfrac{3}{20} + \dfrac{7}{30}$

4. Add.

   a. $1\tfrac{7}{8}$
   $+3\tfrac{5}{6}$

   b. $2\tfrac{3}{4}$
   $+\tfrac{1}{3}$

5. Use a list of multiples to find the LCM of the numbers.

   a. 20 and 25

   b. 4, 6, and 3

REVIEW TEST 10

Name_____

1. Do the division and write your remainder as a common fraction.

   a. $12 \overline{)1,723}$

   b. $68 \overline{)541,591}$

2. Find the LCM of the numbers.

   a. 42 and 60

   b. 18, 12, 15

3. Rewrite $\dfrac{7}{15}$ as a new fraction with the denominator specified.

   a. 60

   b. 45

4. Subtract.

   a. $\dfrac{7}{8} - \dfrac{5}{12}$

   b. $\dfrac{11}{24} - \dfrac{5}{18}$

5. Subtract.

   a. $5\tfrac{2}{5}$
      $-4\tfrac{3}{8}$

   b. $3\tfrac{1}{3}$
      $-1\tfrac{3}{4}$

Review Tests

REVIEW TEST 11

Name _____

1. Write each number below as a product of primes.

   a. 120   b. 780

2. a. Add.

   4,583
   398
   71,864
   2,589
   +55,253

   b. Subtract.

   80,023
   −54,789

3. a. Add.

   $\dfrac{5}{12} + \dfrac{7}{16}$

   b. Add.

   $3\tfrac{1}{4} + 4\tfrac{1}{5}$

4. a. Subtract.

   $\dfrac{7}{10} - \dfrac{2}{15}$

   b. Subtract.

   $9\tfrac{3}{8} - 8\tfrac{5}{6}$

5. a. Use prime factoring to find the LCD of the fractions $\dfrac{5}{27}$ and $\dfrac{7}{45}$.

   b. Change each fraction to a new fraction with the LCD as the denominator and add the fractions.

Review Tests   203

REVIEW TEST 12

Name_____

1. Find the products.

   a. $\dfrac{11}{8} \times \dfrac{4}{5}$

   b. $3\tfrac{1}{5} \times 5\tfrac{5}{8}$

2. Find the sums.

   a. $\dfrac{13}{18} + \dfrac{11}{12}$

   b. $5\tfrac{2}{3}$
      $+2\tfrac{5}{8}$

3. Find the differences.

   a. $\dfrac{7}{12}$
      $-\dfrac{4}{9}$

   b. $10\tfrac{1}{2}$
      $-3\tfrac{7}{10}$

4. Find the products.

   a. 3,418
      × 609

   b. 372
      × 1,000

5. Find the second of two factors.

   a. 8 is the first factor and 56 is their product.

   b. 35 is the first factor and 7,245 is their product.

Review Tests 205

REVIEW TEST 13

Name_____

1. Find the quotients.

   a. $\dfrac{16}{25} \div \dfrac{27}{35}$

   b. $1\frac{3}{10} \div \frac{1}{5}$

2. Find the products.

   a. $\dfrac{54}{35} \times \dfrac{25}{36}$

   b. $1\frac{3}{4} \times 2\frac{4}{7}$

3. Find the answer.

   a. $\dfrac{27}{35} \div \dfrac{18}{45} \times \dfrac{49}{72}$

   b. $2\frac{2}{3} \times \frac{5}{6} \div 1\frac{1}{9}$

4. Reduce using prime factoring.

   a. $\dfrac{90}{150}$

   b. $\dfrac{195}{330}$

5. a. Write 220 as a product of primes.

   b. Find the product of $2 \times 3 \times 3 \times 5 \times 7$.

Review Tests

REVIEW TEST 14

Name_____

1. Write the decimal fraction for

   a. $\dfrac{3}{8}$

   b. $\dfrac{5}{7}$

2. a. Write 13.052 using verbal notation.

   b. Write 13,052 using verbal notation.

3. a. Write 3.275 using expanded notation.

   b. Write 3,275 using expanded notation.

4. Write the following using positional notation.
   a. Five hundred three thousandths.

   b. Five hundred and three thousandths.

5. a. $3\frac{17}{21} \times 8\frac{2}{5}$

   b. $2\frac{5}{12} \div 5\frac{4}{5}$

Review Tests

209

REVIEW TEST 15

Name_____

1. Find the sums.
   a. $6.12 + 0.57 + 8.417 + 867.3$
   b. $0.3 + 0.00575 + 0.0679 + 2.45$

2. Find the differences.
   a. 742.62 and 79.685
   b. $400 - 82.25$

3. Round off 873.475 to the nearest
   a. hundredth.

   b. hundred.

4. Find the sum and round off the answer to the nearest hundredth.
   $15.3527 + 17.26 + 18.375$

5. a. $4\frac{2}{3} \times 2\frac{5}{8}$
   b. $\frac{32}{45} \div \frac{24}{35}$

REVIEW TEST 16

Name _____

1. Find the sums.
   a. $253 + 26 + 2,489 + 7 + 789$
   b. $5.314 + 460.8 + 325.26 + 78.013 + 6.219$

2. Find the differences.
   a. $14.37 - 12.685$
   b. $23 - 4.78$

3. Find the products.
   a. 37×6.9
   b. 14.7×0.259

4. Find the products.
   a. 0.75×525
   b. 0.0396×1.387

5. Find the following products using a shortcut.
   a. $1,000 \times 0.0358$
   b. 100×13.7

REVIEW TEST 17

Name_____

1. Divide.

   a. $0.003 \overline{) 16.35}$

   b. $1.2 \overline{) 1,728}$

2. Find the quotients to the nearest hundredth.

   a. $6.5 \overline{) 102.43}$

   b. $1234 \overline{) 4321}$

3. Find the quotients using a shortcut.

   a. $\dfrac{16.35}{1,000}$

   b. $\dfrac{0.2}{100}$

4. Find the products.

   a. $10,000 \times 0.25$

   b. 1.64×0.09

5. Change each common fraction to a decimal fraction.

   a. $\dfrac{5}{8}$

   b. $\dfrac{1}{6}$

Review Tests

REVIEW TEST 18

Name_____

1. Change to a percent.

    a. 0.75

    b. $\dfrac{3}{8}$

2. Change to a common fraction.

    a. 45%

    b. 0.75

3. Change to a decimal fraction.

    a. 38%

    b. $\dfrac{1}{3}$

4. Change to a decimal fraction.

    a. 225%

    b. $\dfrac{1}{4}$%

5. Use a shortcut to move the decimal point.

    a. 100×0.45

    b. $\dfrac{2.5}{100}$

INDEX

A

Adding Fractions With Different Denominators   93
Addition
    Carrying   11, 12, 14, 17
    Decimal Fractions   145, 151
    Grouping by 10s   13, 14, 17
    Low Stress Algorithm   12, 14, 16
    Of Fractions   83
    Sum   11, 12
    Whole Numbers   11, 14
Algorithm   41, 42, 46
Austrian Method of Subtraction   22, 23

B

Basic Principle of Fractions   73, 76, 78, 79, 88, 89, 93, 116, 126, 165
Borrowing   19, 20, 21, 22, 23

C

Cancelling   116, 117, 119, 120, 121, 122, 125, 128, 129, 130
Carrying. *See* Addition: Carrying
Common Denominator   73, 74, 75, 78
Complex Fraction   126
Composite Numbers   51, 52, 53, 56, 57

D

Decimal Fractions
    Addition   145, 151
    Changing to Common Fractions   174, 175, 180, 181
    Changing to Percents   174, 175, 176, 180, 181
    Division   163, 164, 165, 168, 170
    Multiplication   155, 157, 159
    Repeating   134, 135, 139
    Rounding Off   148, 149, 150, 151
    Subtraction   145, 146, 151
    Terminating   134, 135, 139
Decimal Numeral System   1
Denominator   62, 63, 65, 66, 67, 68
Difference   19, 21, 23
Dividend   42, 43, 44, 45, 46, 163, 164, 167, 168
Dividing Fractions Using Prime Factoring   130

Dividing Mixed Numbers   127, 130
Dividing Mixed Numbers and Fractions   130
Divisibility Rules   53, 54, 56
Division
    Algorithm   46
    By Zero   45, 46
    Decimal Fractions   163, 164, 165, 168, 170
    Dividend   42, 43, 44, 45, 46, 163, 164, 167, 168
    Divisor   42, 44, 46, 164, 167, 168
    Fractions   125, 127, 128, 129, 130
    Inverse   41, 46
    Partial Remainder   43, 46
    Powers of 10   167, 168
    Quotient   41, 44, 46
    Rules   53, 54, 56
    Whole Numbers   41
Division Algorithm. *See* Division: Algorithm
Division By Zero. *See* Division: By Zero
Division of Whole Numbers. *See* Division: Whole Numbers
Divisor   42, 44, 46, 164, 167, 168

E

Expanded Notation   2, 3, 5, 6, 7, 137, 140

F

Factoring
    Prime   105, 106, 108, 110, 118, 119, 120, 121, 122
Factoring Primes   51, 52, 53, 54, 56
Factors   29, 30, 32, 35, 36
    Reversing The Order of While Multiplying   29
Fractions
    Basic Principle   73, 76, 78, 79, 88, 89, 93, 116, 165
    Cancelling   116, 117, 119, 120, 121, 122, 125, 128, 130
    Changing Common to Decimal   174, 180, 181
    Changing Decimal to Common   174, 175, 180, 181
    Changing Decimal to Percent   174, 175, 176, 180, 181
    Common   61, 62, 63, 64, 65, 66, 67, 68
        Improper   63, 64, 65, 68
        Proper   63, 64, 68, 69
    Common Denominator   73, 74, 75, 78
    Decimal   61, 133, 134, 135, 137, 138, 139, 140
    Denominator   62, 63, 65, 66, 67, 68
    Division   125, 127, 128, 129, 130
    Improper   63, 64, 65, 68, 69
    Inverting   125, 126, 127, 128, 129, 130
    LCD (Lowest Common Denominator)   105, 106, 108, 110, 111

    LCM (Least Common Multiple)   87, 88, 89, 90, 92, 93
    Lowest Terms   65, 66, 67, 68
    Multiplying   122
    Numerator   62, 63, 65, 66, 67, 68
    Proper   63, 64, 69
    Reducing   64, 66, 68
    Repeating Decimal   134, 135, 139
    Subtraction   97, 101
    Terminating Decimal   134, 135, 139

G

Grouping by 10s. *See* Addition: Grouping by Tens

I

Inverse   19, 41, 46, 125
Inverting   125, 126, 127, 128, 129, 130

L

LCD   105, 106, 108, 110, 111
LCM   87, 88, 89, 90, 92, 93
Least Common Denominator   75
Least Common Multiple (LCM). *See* LCM
Low Stress Algorithm. *See* Addition: Low Stress Algorithm
Lowest Terms   65, 66, 67, 68

M

Mixed Numbers   61, 64, 65, 68, 69, 70
Mixed Numbers Versus Improper Fractions   93, 101
Multiples   86, 87, 88, 93
Multiplication
    By 10, 100, 1000, ...   34
    Decimals   155, 157, 159
    Factors   29, 30, 32, 35, 36
    Fractions   122
    Perfect Squares   30, 31, 36
    Powers of 10   158, 159
    Product   29, 30, 31, 32, 33, 36, 37, 39
    Shortcuts Involving Zeros   36
    Table   29, 30
    Whole Numbers   29, 36
Multiplicative Inverse. *See* Reciprocal
Multiplying Mixed Numbers With Common Fractions   122
Multiplying More Than Two Fractions Together   122
Multiplying Whole Numbers With Fractions   122

N

Number   1, 2, 5, 6
Numeral   1, 2, 3, 4, 5, 7, 8
Numerator   62, 63, 65, 66, 67, 68

O

Order of Operations   128, 130

P

Partial Remainder   43, 46
Percent   173
Percent Symbol. *See* Percents: Symbol
Percents
    Changing to Common Fractions   178, 180, 181
    Changing to Decimal Fractions   177, 178, 180, 181
    Symbol   173
Perfect Squares   30, 31, 36
Place Value   2, 3, 4, 5, 7, 136, 137, 138, 139
Place Value Chart   137
Positional Chart   7
Positional Notation   1, 3, 4, 5, 6, 134, 138, 141
Powers of 10   3, 5, 158, 159, 167, 168
Prime Factor-
    ing   51, 52, 53, 54, 56, 105, 106, 108, 110, 118, 119, 120, 121, 122
Prime Numbers   51, 52, 54, 55, 56
Product   29, 30, 31, 32, 33, 36, 37, 39
Proper Fractions   63, 64, 69

Q

Quotient   41, 44, 46

R

Reciprocal   125
Reducing Fractions   64, 66, 68
Relative Primes   108, 109, 110
Rounding Off   148, 149, 150, 151
Rule For Dividing Fractions   127

S

Squaring A Number   31, 36
Subtracting Mixed Numbers   101
Subtraction
    Austrian Method   22, 23
    Borrowing. *See* Borrowing

    Checking   21, 23
    Decimal Fractions   145, 146, 151
    Difference. *See* Difference
    Fractions   97, 101
    Whole Numbers   19, 23
Sum. *See* Addition: Sum

T

Trial Divisor   42, 44, 45, 46

V

Verbal Notation   2, 3, 4, 5, 7, 8, 138, 141

W

Whole Numbers
    Addition   11, 12, 14
    Commas   4, 8
    Division   41
    Multiplication   29, 36

Factor into prime

12600
100 × 126
4 × 25   2 × 63
2×2 5×5 7 × 9
          3 × 3

$2^3 \times 3^2 \times 5^2 \times 7$

135
5 × 27
3 × 9
3 × 3

$3^3 \times 5$

① Factor into prime

375
5 × 75
3 × 25
5 × 5

```
      75
  5)375
    -35↓
      25
     -25
       0
```

```
     25
  3)75
     6↓
     15
    -15
      0
```

$5^3 \times 3$

② 196
2 × 98
2 × 49
7 × 7

```
     98
  2)196
    18↓
     16
     16
      0
```

```
    49
  2)98
    8↓
    18
   -18
     0
```

$2^2 \times 7^2$

③ Divide

```
           6022
       7)42,156
         42↓
          01
           0↓
           15
           14↓
            16
           -14
             2
```

6022 R.2

```
      6022
    ×    7
    42154
    +    2
    42156
```
one at a time only

④
```
       572
   16)9154
      -80↓
       115
       112↓
        34
        32
         2
```

```
   416
  ×  7
   112 (2912?)
```

572 R.2

```
     10200
  6)61200
    -6↓
     01
      0↓
      12
     12
      00
       00
       00
       0
```

⑤ Multiply
```
        914
      ×   ½
      1652 (?)
      1828
     4570 0
     548400
     595928
```

⑥
```
        8,600
      ×   195
         
         195
          86
        1170
        1560
    1,677,000
```
Answer Ends in 2 0's. Add them at the end so

Subtract
```
        5982
      - 67943
        48039
```

addition
```
   94 + 115 + 608 + 39
   608
   115
    94
  + 39
   856
```

⑨ Write in verbal notation
903,640,875
nine hundred three million, six hundred forty thousand, eight hundred seventy five.

⑩ "write in positional notation"
Ninety million seventy two thousand five hundred six

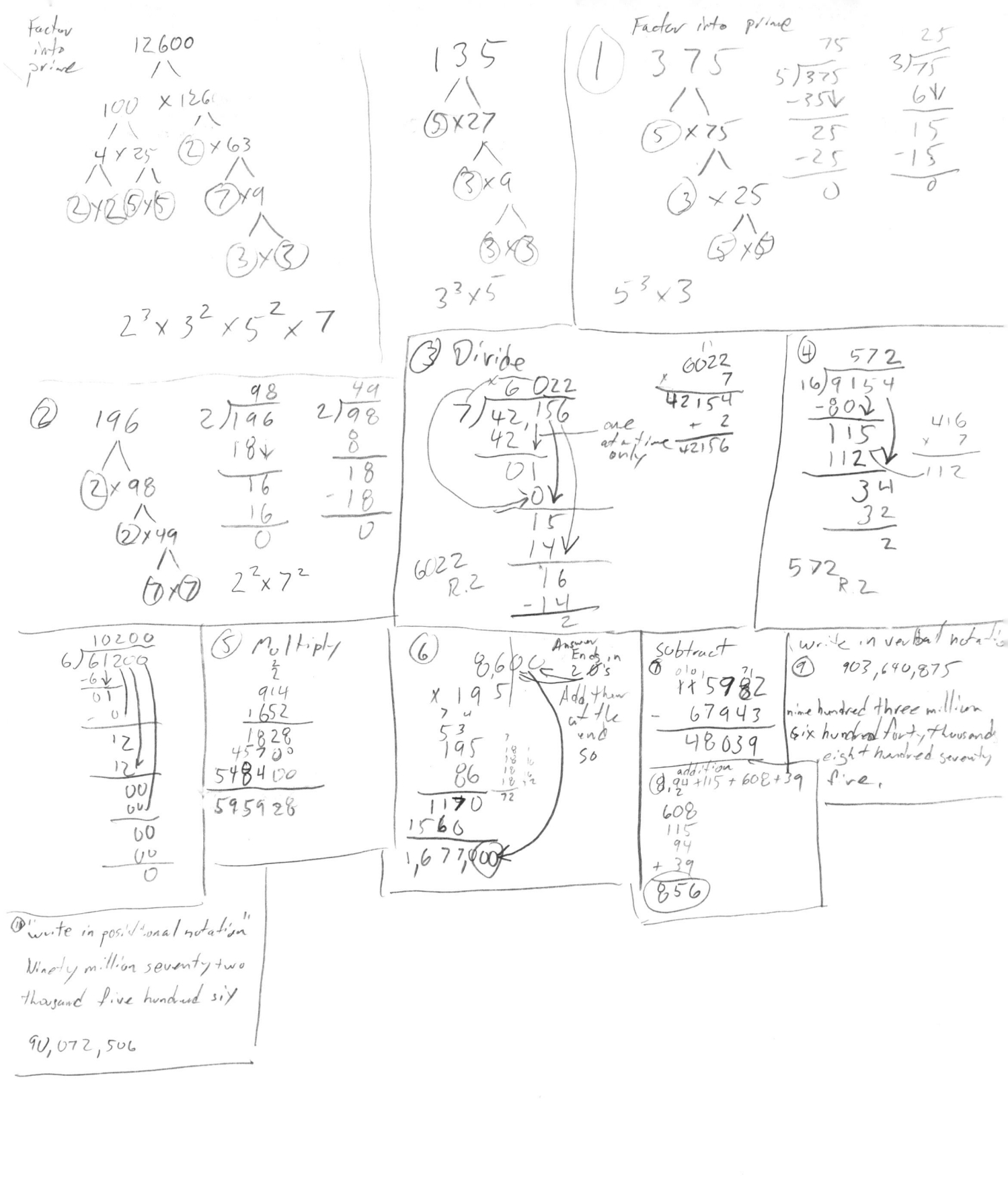

90,072,506